MW01482282

Canadian
Energy Efficiency Outlook

A National Effort
For Tackling Climate Change

Canadian
Energy Efficiency Outlook

A National Effort
For Tackling Climate Change

Pierre Langlois
Geneviève Gauthier

THE FAIRMONT PRESS, INC.

CRC Press
Taylor & Francis Group

Library of Congress Cataloging-in-Publication Data

Names: Langlois, Pierre (Writer on energy efficiency), author. | Gauthier, Geneviève, author.
Title: Canadian energy efficiency outlook : a national effort for tackling climate change / Pierre Langlois, Geneviève Gauthier.
Description: 1 Edition. | Lilburn, GA : Fairmont Press, Inc., [2018] | Includes bibliographical references and index.
Identifiers: LCCN 2018015672 (print) | LCCN 2018028453 (ebook) | ISBN 0881737844 (Electronic) | ISBN 9780881737844 (Electronic) | ISBN 0881737836 (alk. paper) | ISBN 9781138311336 (Taylor & Francis distribution : alk. paper)
Subjects: LCSH: Power resources--Canada--Forecasting. | Energy policy--Canada.
Classification: LCC HD9502.C22 (ebook) | LCC HD9502.C22 L36 2018 (print) | DDC 333.790971/01--dc23
LC record available at https://lccn.loc.gov/2018015672

Published by The Fairmont Press, Inc.
700 Indian Trail
Lilburn, GA 30047
tel: 770-925-9388; fax: 770-381-9865
http://www.fairmontpress.com

Distributed by Taylor & Francis Group LLC
6000 Broken Sound Parkway NW, Suite 300
Boca Raton, FL 33487, USA
E-mail: orders@crcpress.com

Distributed by Taylor & Francis Group LLC
23-25 Blades Court
Deodar Road
London SW15 2NU, UK
E-mail: uk.tandf@thomsonpublishingservices.co.uk

Printed in the United States of America
10 9 8 7 6 5 4 3 2 1

10:0881737836   (The Fairmont Press, Inc.)
13:9781138311336   (Taylor & Francis Group LLC)

While every effort is made to provide dependable information, the publisher, authors, and editors cannot be held responsible for any errors or omissions.

The views expressed herein do not necessarily reflect those of the publisher.

Table of Contents

PART I—Federal, Provincial and Territorial Reviews 1

Chapter 1
Federal 3
  Energy Sector 3
  Energy Efficiency Background .4
  Stakeholders 7
  Policies and Strategies 10
  Training and Skills Development 12
  Legal and Regulatory Framework 13
  Financial Mechanisms 19

Chapter 2
Alberta 23
  Energy Sector 23
  Energy Efficiency Background 24
  Stakeholders 26
  Policies and Strategies 28
  Legal and Regulatory Framework 28
  Financial Mechanisms 29

Chapter 3
British Columbia 31
  Energy Efficiency Background 31
  Stakeholders 33
  Policies and Strategies 35
  Legal and Regulatory Framework 41
  Financial Mechanisms 46

Chapter 4
Manitoba 51
  Energy Sector 51
  Energy Efficiency Background 51

Stakeholders 56
Policies and Strategies 57
Legal and Regulatory Framework. 58
Financial Mechanisms 60

Chapter 5
New Brunswick. 63
Energy Sector 63
Energy Efficiency Background. 64
Stakeholders 67
Policies and Strategies 69
Legal and Regulatory Framework. 71
Financial Mechanisms 72

Chapter 6
Newfoundland and Labrador 73
Energy Sector 73
Energy Efficiency Background. 73
Stakeholders 78
Policies and Strategies 80
Legal and Regulatory Framework. 84
Financial Mechanisms 85

Chapter 7
Northwest Territories 91
Energy Sector 91
Energy Efficiency Background. 93
Stakeholders 97
Policies and Strategies 98
Legal and Regulatory Framework. 101
Financial Mechanisms 102

Chapter 8
Nova Scotia 105
Energy Sector 105
Energy Efficiency Background. 105
Stakeholders 112
Policies and Strategies 114

Legal and Regulatory Framework. 117
Financial Mechanisms 120

Chapter 9
Nunavut **123**
Energy Efficiency Background. 123
Stakeholders 126
Policies and Strategies 127
Legal and Regulatory Framework. 129
Financial Mechanisms 130

Chapter 10
Ontario **131**
Section I—Electricity Sector 131
Section II—Natural Gas Sector. 166

Chapter 11
Prince Edward Island **173**
Energy Sector 173
Energy Efficiency Background. 173
Stakeholders 178
Policies and Strategies 180
Legal and Regulatory Framework. 181
Financial Mechanisms 182

Chapter 12
Quebec **183**
Energy Efficiency Background. 183
Stakeholders 189
Policies and Strategies 197
Legal and Regulatory Framework. 200
Financial Mechanisms 202

Chapter 13
Saskatchewan **209**
Energy Sector 209
Energy Efficiency Background. 209
Stakeholders 213
Policies and Strategies 216

Legal and Regulatory Framework. 217
Financial Mechanisms 218

Chapter 14
Yukon. **221**
Energy Efficiency Background. 221
Stakeholders 221
Policies and Strategies 222
Legal and Regulatory Framework. 223
Financial Mechanisms 223

PART II—Best Practices and Case Studies **227**

Chapter 15
Energy Performance Contracts (EPCs) **229**
Description. 229
History 230
Market 231
Policies, Programs and Legal Framework 232
Conclusions 238

Chapter 16
Program Evaluation **241**
What is EE Program Evaluation? 241
Approaches 242
Specific Applications 243
Level of Investment. 243
Market Actors 243

Chapter 17
Capacity Building and Training **245**
Capacity Building and Training Activities 246
Main Stakeholders. 247
Examples 248

Chapter 18
The Canadian Energy Efficiency Alliance (CEEA) **253**

Introduction 253
The Origins of CEEA 254
CEEA Version 2.0 255
The Move to CEEA Version 3.0 256

Chapter 19
EfficiencyOne: A Unique Implementation Model 259
Introduction 259
Results 260
Services. 262
Pillars of Success 262
Conclusion. 264

Chapter 20
Retscreen: Clean Energy Management Software 265
Background 265
Stakeholders 266
Features. 267

Chapter 21
The Atmospheric Fund 269
A Municipal Climate Pioneer 269
Early Investments 269
Approach. 270
Priority Focus 271
The Path Ahead 272

CONCLUSION 273

APPENDICES 277
Authors. 277
Contributing Authors. 280
Glossary of Acronyms 287
References by Chapter 294

INDEX 321

Foreword

Energy efficiency (EE) has been recognized since the early 1970s as the most relevant mechanism to optimize the way we meet our energy needs. Not only is it financially viable for end users, but it enables utilities to optimize their financial returns and the use of their infrastructure. For the last 25 years, EE has also been identified as the favored approach to fighting climate change. It is recognized as the most cost-efficient means of achieving significant results in carbon emission reductions in the shortest time possible.

Canada has been a champion of EE since it started being recognized as such a unique mechanism. Over the last 40 years, Canadian public organizations at all levels have led the field in the introduction of policies, institutional frameworks, programs and innovative implementation schemes. The introduction of EE building codes, standards and labels, as well as energy performance contracting schemes are but a few initiatives developed in Canada over these 40 years.

The EE Canadian experience is one that involves many government actors (federal, provincial, municipal), nonprofit organizations and private sector firms that all contribute to the successful implementation of EE. Learning about the historical development of EE policies and initiatives, and where we now stand at the federal and provincial levels, enables us to understand how we have achieved so many successes.

EE is now part of our lives. Although much has been accomplished in the last 40 years, so much more remains to be done to benefit fully from its potential. This is why more and more actors from all sectors throughout Canada are currently committing to promoting and using EE. These efforts have been led by our governments at all levels. They have developed and improved the legal and institutional frameworks to increase the use of EE in the energy mix of the country. Many other actors such as utilities now fully integrate EE in their operations. The political structure of our country and the constant evolution of the EE sector, especially in the last 10 years, as well as the greatly varied environments in which EE is utilized make it difficult to gain clear understanding of how governments and stakeholders operate and support the development of the EE field.

The rationale behind this book is to present where the Canadian EE sector stands today to all Canadian stakeholders and those interested around the world. The Canada Energy Efficiency Outlook aims to outline the different environments that support EE development in our highly diversified provinces and territories, as well as at the national level, and consequently allow the reader to better understand the complexities involved. More globally, this book serves as an important reference for all interested parties on how Canada has variably innovated and developed mechanisms to achieve the goal of making our country more energy efficient.

Geneviève Gauthier
Pierre Langlois

Acknowledgments

Authoring such a book as the *Canadian Energy Efficiency Outlook* was a major challenge from the beginning. Indeed, Canada is a vast and diversified country where the energy sector is a provincial jurisdiction, meaning there are as many legal and regulatory frameworks as there are provinces and territories.

We launched this project more than a year ago and knew at the time that it could not be achieved without the support of many stakeholders who hold the knowledge of where things stand and are developing around the country. When we started reaching out to these people, we received instant and continued support from them throughout the whole process of developing this book.

Furthermore, over the course of this work, we relied on many great people with the talent and the will to make this project a success. We could not have completed it without them.

We therefore thank all the contributors and supporters who made the publication of this book possible.

Let us start with all the coauthors whose bios are presented in the appendix. These coauthors constitute the main actors in the Canadian EE sector. Their exceptional knowledge of their respective provincial and territorial environments truly enabled us to include the most current information on the Canadian EE landscape.

The coauthors and other experts who helped us develop the Best Practices and Case Studies section herein also merit our thanks due to the quality of their contributions and the knowledge they provided. They greatly contribute to developing the best initiatives in the EE sector.

We also thank all Econoler staff involved in the preparation of this book: Jean-François Bergeron, Jayson Bérubé, Amandine Gal, Amandine Lanneval, Paola Mendez, Lorenz Mootoosamy, France Parent and Denitsa Ruseva. They supported us in the research, as well as reviews, proofreading and editing of all the chapters that constitute this work. Their professionalism and dedication were key to the realization of this book.

Last but not least, we thank the great team at the Fairmont Press with whom we have had the opportunity of collaborating on other publications. They have always pushed us to provide our best by positively reviewing and critiquing the work produced. Only such a structured and intensive process can ensure that a publication exemplifies the high standard established by the Fairmont Press.

Introduction

Canada is one of the most significant energy consuming nations per capita in the world. The climate, size and population distribution of Canada make optimizing the use of energy very challenging. In such a context, EE has unsurprisingly been identified as an essential component of the energy mix.

Over the last 40 years, the federal and provincial governments have developed legal and institutional frameworks to introduce and enable the growth of EE in their respective jurisdictions. They have also supported the development of all kinds of initiatives and programs with the aim of transforming the energy market and making it more efficient.

When asked about how Canada evolved over the years and what is the current status of EE in Canada, most experts are puzzled by the question and find it difficult to provide a befitting answer. Indeed each province and territory has developed its own approach to fostering EE, and they are all at different implementation stages. The models used and actors involved differ greatly. Also, the federal government has developed many of its own initiatives and worked in collaboration with many stakeholders to provide direction and support to enable the country and market actors to become, as a whole, more energy efficient.

The Canada Energy Efficiency Outlook aims to answer the above question by presenting the history and current state of the EE sector all around Canada. Therefore, this book first presents the national context and the role of key federal market actors. It then presents each province's and territory's history with the EE sector and how legal and institutional frameworks have evolved. We tried here to standardize the presentation so the information can be more easily compared and contrasted. Finally, the book culminates with a selection of case studies that both underscore those specific aspects of the EE sector that might be of national interest and showcase unique and innovative initiatives in the Canadian context.

We hope that the *Canada Energy Efficiency Outlook* is of interest to all Canadian EE stakeholders and that it imparts understanding of both how the EE sector developed and evolved over the years in different markets, as well as the current status of EE in each province and territory. We have also sought for the contents of this book to be of interest to stakeholders around the world, imparting more knowledge about the world-leading Canadian EE market. We have further endeavored to ensure readers draw lessons learned and best practices developed by Canada. Finally, we hope that this publication contributes to the continuous development of the EE market in Canada and abroad.

The main authors,

Geneviève Gauthier
Pierre Langlois

Part I

Federal, Provincial and Territorial Reviews

Chapter 1

Federal

ENERGY SECTOR

Canada's energy consumption has been steadily increasing since 1995. Annual average consumption between 2013 and 2015 attained 195,338 ktoe, a 19% increase over annual consumption in the 1995-97 period. The transportation and industrial sectors consumed the most, each accounting for 33% of total consumption in 2013-15. Energy consumption in both sectors grew more rapidly than total consumption at 25 and 27% respectively for the 20-year period from 1995 to 2015. The agriculture sector increased consumption by 27% over the same 20-year period, accounting for 3% of total consumption in 2013-15.

The residential and commercial sectors both increased consumption by slightly more than 8% between 1995 and 2015. In 2013-15, the residential sector consumed 17% of total consumption, while the commercial sector consumed 12%. The public administration sector remained stable throughout the 20-year period, representing 2% of Canada's total consumption in 2013-15.

Figure 1-1 illustrates average energy consumption in three-year periods from 1995 to 2015.[1]

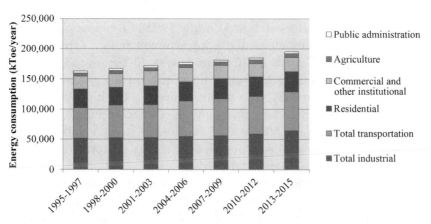

Figure 1-1: Canada Energy Consumption from 1995 to 2015

ENERGY EFFICIENCY BACKGROUND

The energy sector is important to the Canadian economy. Canada is one of the largest energy exporters in the world and the economy relies on the energy sector and the employment created thereof. Additionally, energy ensures the well-being of Canadians given the challenging climate and low population density of 3.5 people/km^2, among the lowest in the world. These factors increase the need for heating and transportation which, in turn, increase energy requirements.

Despite the abundance of energy resources, Canada was affected by the oil peak prices of the 1970s and 1980s, and the government responded by introducing the first EE initiatives.

Thus, in 1975 the first program targeting the efficient use of energy was launched, the Canadian Industry Program for Energy Conservation (CIPEC). This program, a voluntary partnership between private and public stakeholders, aimed to generate effective actions that reduce industrial energy use per unit of production. Currently, more than 1,400 companies voluntarily participate in CIPEC.[2]

CIPEC was followed by another voluntary but innovative mechanism, EnerGuide. This comparative label displays basic energy information on appliances to permit consumers to compare and consider energy consumption when making buying decisions. Introduced in 1978 as a voluntary initiative, EnerGuide became mandatory for major household appliances in 1995 with the advent of the Energy Efficiency Act. Since then, EnerGuide has been an important pillar of Canadian EE policies.

After energy prices stabilized in the 1980s, EE policies and measures persisted as key elements in government policy, especially in Natural Resources Canada (NRCan) initiatives. EE became a key mechanism in support of Canada's commitment to reducing greenhouse gas (GHG) emissions. In 1981, EE was introduced in the buildings sector through the R-2000 Standard program. The Canadian Home Builders' Association (CHBA) and NRCan designed and implemented this voluntary program whose goal is to improve the energy performance of new homes by establishing higher standards.

In 1992, and amidst debate over international environmental concerns, Canada developed its first EE law, the Energy Efficiency Act. This Act empowered the Minister of Natural Resources to promote the efficient use of energy and alternate fuel sources.[5] The law mandated NR-

There are two discernable types of energy labels, endorsement labels and comparative labels.

Endorsement labels are a seal of approval indicating that a product meets certain specified energy consumption criteria. Their purpose is to clearly indicate to the consumer that a given product saves energy compared to others that do not achieve minimum requirements. These labels are generally based on the yes or no principle (i.e. they indicate that a product uses more or less energy than a specified threshold), and they offer little additional information because they serve to provide simple but specific information that customers can easily use. Typically, endorsement labels apply to the top tier (i.e. the top 15 to 25%) energy efficient products in a market. One example of an EE endorsement label is the U.S. ENERGY STAR® label.

Comparative labels allow consumers to compare energy use among available models to make informed purchasing choices. Generally, two forms of comparative labels are in use around the world; one uses a categorical ranking system, while the other uses a continuous scale or bar graph to illustrate relative energy use. A third form, the information-only label, provides information about the labeled product without comparing energy use to other models. Information-only labels are not often used for promoting EE.[3], [4]

Can to design, implement and enforce minimum energy performance standards (MEPS), introduce labeling and collect data on the energy consumption of energy-using products. Three years later, the Energy Efficiency Regulations (the Regulations) were introduced. These Regulations defined EE standards for commercial energy-using products. They also prescribed labeling requirements for certain products, such as disclosing the energy use of a given product and comparing it to others in the same category.[6]

During the '90s, Canada introduced several initiatives aimed at promoting EE across many sectors, such as the Federal Buildings Initiative (FBI) and National Energy Use Database (NEUD). The FBI provided knowledge, training and expertise to support EE retrofit projects in buildings owned by federal government organizations, whereas the NEUD aimed at providing unique, reliable, Canada-wide information on energy consumption across all sectors over the long term by centralizing more detailed energy information required to make informed decisions.

Access to information and knowledge has always been an important pillar of Canadian EE strategies. Several initiatives were introduced to increase EE awareness in all sectors. For instance, the Dollars to $ense Energy Management Workshops were introduced in 1997 and are still being delivered. So far, the program has trained more than 30,000 representatives from industrial, commercial and institutional organizations.

Whereas several programs and initiatives from the '70s and '80s are still in place, other programs have immerged in recent years in accordance with changes in technologies and newly available research. Standards have been continuously updated (new labeling requirements, new standards for appliances and homes), and new financing schemes have been developed. Now, Canadians can not only identify energy efficient homes, but also have access to a variety of finance conditions through the Green Home Program which provides up to a 25% refund to borrowers when purchasing either an energy efficient home or an existing home and making EE improvements. Businesses can also invest in EE technologies thanks to an incentive provided since 1994 through the Capital Cost Allowance (CCA).

The federal government has also supported the efficient use of resources in the transport sector by introducing initiatives targeting personal vehicles, as well as commercial and institutional fleets. For example, NRCan has administered the SmartWay Program since 2012, a government-industry collaboration that tracks and benchmarks the performance of commercial vehicles. FleetSmart is another federal program aimed at providing tools and information to help reduce fuel consumption. Furthermore, the federal government and NRCan have set EE standards for motor vehicles and implemented labels for new vehicles through the EnerGuide initiative.

Internationally, Canada has played an active role in promoting EE. In 2009, Canada, in conjunction with G8 members, founded the International Partnership for Energy Efficiency Cooperation (IPEEC) to promote collaboration on EE. In 2010, Canada joined the Clean Energy Ministerial initiative under which activities such as the Global Superior Energy Performance Partnership[i] and the Super-efficient Equipment and Appliance Deployment[ii] are carried out.

---

[i] This initiative focuses on energy management systems for industry (i.e. ISO 50001).

[ii] This initiative seeks to accelerate market transformation toward high-efficiency equipment and appliances through standards and labels.

Canada has also very actively promoted activities developed under the Energy and Climate Partnership of the Americas (ECPA). This initiative was launched in 2009 when leaders in the Western Hemisphere reaffirmed their commitment to working together for a clean energy future. A main pillar of ECPA is promoting the development of EE policies through cooperation and exchange.

Canada has a long national and international track record in promoting the efficient use of energy resources. Activities and measures have been developed for all sectors and all stakeholders, whether they be federal, individual or private-sector actors. These efforts have yielded concrete results: it is estimated that, from 1990 to 2013, EE in Canada improved by 24% overall, thus saving Canadians more than $37 billion [iii] on energy bills in 2013 alone.[5]

The following sections present a panorama of EE initiatives and actions conducted by the Government of Canada over the last 20 years to foster the efficient use of energy resources across different sectors.

STAKEHOLDERS

Canada is a federation composed of ten provinces (Alberta, British Columbia, Manitoba, New Brunswick, Newfoundland and Labrador, Nova Scotia, Ontario, Prince Edward Island, Quebec and Saskatchewan) and three territories (Northwest Territories, Nunavut and Yukon).

According to the Canadian Constitution, provinces have the right to enact their own legislation regarding the exploration, conservation and management of non-renewable natural resources. They may also legislate in matters pertaining to natural and forestry resources, as well as the development, conservation and management of sites and facilities for electric energy generation.

Since energy issues fall primarily under provincial jurisdiction, provinces and territories have their own regulatory bodies and energy market structures. For example, Alberta has a fully competitive wholesale and retail electricity market, whereas Nova Scotia has vertically integrated utilities.

Nevertheless, the federal government has the mandate to regulate on matters that involve cross-border energy trading and infrastructure, as well as the management of offshore energy resources.

[iii] All dollar amounts herein are expressed in Canadian currency (CAD) unless otherwise indicated.

NRCan is the lead department on energy policy at federal level. NRCan was created through the Department of Natural Resources Act of 1994. The NRCan Office of Energy Efficiency (OEE) is the primary body responsible for EE policy and programs. It was formed in 1996/97 as the immediate successor to the Efficiency and Alternative Energy Branch. The OEE is under the umbrella of the NRCan Energy Sector Unit and advises the government on federal energy policies and strategies.

The OEE administers EE regulations and manages EE programs such as the NEUD. It also offers grants, incentives and other resources such as workshops and publications aimed at increasing awareness and knowledge on EE issues.[5]

As part of NRCan, the Office of Energy Research and Development (OERD) coordinates energy research and development activities for the Government of Canada and acts as liaison with the International Energy Agency. The OERD is responsible for various initiatives including the Program of Energy Research and Development (PERD).

Also at NRCan, CanmetENERGY has been contributing to Canada's energy development for more than 100 years. In the 1970s, the Mines Branch of the Fuels and Fuels Testing Division was renamed CANMET (the Canada Centre for Mineral and Energy Technology), which was later divided in two groups: energy and minerals.

Today, CanmetENERGY focuses on the development of cleaner fossil fuels and related environmental technologies with a focus on oil sands and heavy oil. It also conducts research and development in the areas of EE as well as renewable and alternative energy sources. CanmetENERGY also leads programs related to buildings, communities, renewables and industrial processes, as well as manages the RETScreen International Clean Energy Decision Support Centre.

The Steering Committee on Energy Efficiency (SCEE) is the primary mechanism for cooperation and engagement between federal, provincial and territorial governments on EE and alternative energy fuel issues. The SCEE was established in 2004 by the Council of Energy Ministers which comprises federal, provincial and territorial energy ministers. SCEE provides an annual report to the Energy and Mines Ministers' Conference (EMMC).

EE activities need to be coordinated and implemented in close coordination and collaboration between federal government departments, the private sector and civil society to ensure success. For this reason, a single federal department cannot develop, implement and monitor all

EE activities on its own. In fact, the soundness of Canada's approach to EE is manifest in the countless successful examples of intergovernmental collaborations. Specifically, the Canada Border Services Agency (CBSA)[iv] assists NRCan in administering both the Energy Efficiency Act and the Energy Efficiency Regulations.[v] NRCan also partnered with the Canadian Home Builders' Association (CHBA)[vi] to develop the R-2000 Standard program. Furthermore, the NRC[vii] established the Canadian Commission on Building and Fire Codes (CCBFC)[viii] which develops and updates the National Energy Building Code of Canada for Buildings (NECB). More recently under the auspices of the NEUD, NRCan partnered with Statistics Canada (StatCan)[ix] to conduct the 2011 Survey of Household Energy Use (SHEU-2011).

Other federal agencies and departments develop and implement EE activities as well. For instance, Transport Canada[x] runs initiatives to support the development of new and more efficient transport vehicles. Canada Mortgage and Housing Corporation (CMHC)[xi] promotes access to new energy efficient homes or facilitates financing for EE home improvements through its Green Home program.[7]

The Canada Revenue Agency (CRA) is responsible for administering Canadian tax laws including the Income Tax Act. Class 43.1 of this Act provides an accelerated capital cost allowance for specified energy efficient and renewable energy (RE) technologies.

[iv]   CBSA is a federal institution that oversees international travel and trade across the Canadian border.

[v]   Importers who are dealers of regulated products must provide the CBSA with prescribed data to be included in the release package transmitted to the CBSA electronically through the Other Government Department Interface and the Single Window Initiative. http://www.cbsa-asfc.gc.ca/publications/dm-md/d19/d19-6-3-eng.html?wbdisable=true.

[vi]   The CHBA is the construction industry association and counts over 8,500 company members.

[vii] NRC is the Government of Canada's research and technology organization.

[viii] The CCBFC is an independent committee of volunteers.

[ix]   StatCan is the federal agency responsible for the collection, compilation and dissemination of official statistics in Canada. StatCan is mandated through the Statistics Act to collect energy information at the national and provincial/territorial levels.

[x]   Transport Canada is a federal institution responsible for transportation policies and programs.

[xi]   CMCH is Canada's housing authority and is mandated to facilitate access to housing and contribute to financial stability by helping Canadians meet their housing needs.

The Federation of Canadian Municipalities (FCM)[xii] funds municipal initiatives including EE programs through the Green Municipal Fund (GMF).

Environment and Climate Change Canada (ECCC) is the lead federal department for a wide range of environmental issues including climate change initiatives. Responsibilities include the implementation of the Pan-Canadian Framework on Clean Growth and Climate Change.

POLICIES AND STRATEGIES

Since natural resources and energy are administered and regulated at the provincial level, each province and territory may define its own energy objectives, activities and strategies. Nevertheless, it has been acknowledged by federal, provincial and municipal leaders that a unified vision and collaborative actions are required for Canada to achieve sustainable development.

This vision was reflected in the first Energy Strategy developed in 2007 by the Council of the Federation (COF)[xiii] in a document entitled *A Shared Vision for Energy in Canada*. This strategy was based on energy plans previously developed by provinces and territories and established a seven-point action plan aimed at striking a balance between secure energy supply, environmental and social responsibility, and continued economic growth and prosperity.

At the 2012 COF meeting, premiers agreed to renew the 2007 Strategy and ensure its continued relevance to the Canadian energy landscape. The Canadian Energy Strategy Working Group was thereby established, which subsequently developed the Canada Energy Strategy in 2015. Under this new strategy, the role of EE was highlighted and three main goals were defined: (1) strengthen Canadians' understanding of the benefits of EE and conservation; (2) maximize access to energy savings for all energy consumers; and (3) encourage market transformation through targeted EE and conservation policies and regulations.

[xii] The Federation of Canadian Municipalities (FCM) is a non-profit organization that represents over 2,000 municipalities.

[xiii] The Council of the Federation was established in 2003, enabling premiers to work collaboratively to strengthen the Canadian federation by fostering a constructive relationship among the provinces and territories and with the federal government. Source: http://www.canadaspremiers.ca/en/about.

Hence, EE is recognized as a key issue in the energy strategy, but it is also a key pillar to achieve international climate change commitments and help Canada transition to and ensure sustainable development for Canadians.

Canada's engagement with the challenges of climate change started in the early '90s. In 1992, after the United Nations Framework Convention on Climate Change treaty was adopted, Canada renewed and expanded its support for EE initiatives. Since then, many GHG plans, strategies and actions have been defined and implemented, such as the 1995 National Action Program on Climate Change (NAPCC), the Action Plan 2000 on Climate Change of 1998 and the Climate Change Plan for Canada of 2002.

In 2007, the federal government introduced the Clean Air Agenda (CAA) as a fundamental component of Canada's efforts to address the challenges of climate change and air pollution.

In March 2016, Canada's First Ministers issued the Vancouver Declaration on Clean Growth and Climate Change. In this Declaration, the prime minister and premiers emphasized their joint commitment to EE and pledged to reduce national GHG emissions by 30% below 2005 levels by 2030 and transition toward a low-carbon and clean growth economy.

On 22 April 2016, Canada was among 175 parties to sign the Paris Agreement on Climate Change. Then in December 2016, the Government of Canada released the Pan-Canadian Framework on Clean Growth and Climate Change, which constitutes Canada's plan to meet emission reduction targets and achieve Canada's international commitments in the Paris Agreement. The Framework outlines critical actions including: (1) develop new building codes to increase EE; (2) make more efficient use of existing power supplies; and (3) drive significant reductions in emissions from government operations.

Central to the Framework is establishing a price on carbon. Some provinces have already introduced carbon pricing systems and the federal government announced in October 2017 that it would apply a minimum price on carbon by 2018 to ensure a minimum price on pollution across the country.

In 2008, the Federal Sustainable Development Act established the requirement to develop the Federal Sustainable Development Strategy (FSDS).[xiv] The purpose of this Act was to provide the legal framework

---

[xiv] Environment and Climate Change Canada plays a key role in implementing the Act.

for developing and implementing the FSDS to render environmental decision making more transparent for and accountable to Parliament.[8]

The FSDS defines sustainable development priorities, sets out specific and measurable targets and milestones, as well as clear action plans to track and monitor targets. The FSDS is aligned with international commitments[xv] and national strategies and, thus, recognizes the importance of EE in all sectors of the economy.

The FSDS 2016-2019 is the main mechanism for sustainable development planning and reporting, and it represents Canada's efforts to implementing the Sustainable Development Goals (SDGs) of the 2030 Agenda for Sustainable Development.

TRAINING AND SKILLS DEVELOPMENT

Training and skills development has been at the center of EE policies and, for this reason, the Government of Canada has always developed programs and actions to increase EE related skills and certification programs such as Dollars to $ense Energy Management Workshops and RETScreen®.

Dollars to $ense
The Dollars to $ense Energy Management Workshops have been in place since 1997 and constitute a component of the CIPEC. These workshops teach industry actors how to improve operational efficiency, create a better work environment and reduce GHG emissions. They also serve to train building designers, builders, owners and operators through building design simulations on new technologies and best practices that improve EE in new and existing buildings.

So far, it is estimated that more than 10,000 representatives from industrial, commercial and institutional organizations have participated in the workshops which have resulted in an estimated annual reduction of 3.32 PJ in energy use.

RETScreen®
NRCan officially released RETScreen in 1998. This software, originally designed for evaluating renewable energy technology projects

---

[xv] In other words, the Sustainable Development Goals (SDGs).

worldwide, was developed by NRCan and a team of more than 40 industry and government experts. Today, RETScreen is a Clean Energy Management Software system used to analyze the feasibility of EE, renewable energy and cogeneration projects. This globally recognized software has been awarded local and international distinctions such as the 2010 Energy Globe and the 2005 NASA Group Achievement Award. RETScreen is currently available in 36 languages and has over 465,000 users across the world.

RETScreen® International is managed by the NRCan CANMET Energy Technology Centre—Varennes (CETC—Varennes).

LEGAL AND REGULATORY FRAMEWORK

Current EE Laws

As previously mentioned, Canada enacted the Energy Efficiency Act in 1992. This Act is the most important piece of EE legislation because it grants the Canadian government the authority to establish MEPS and implement energy labeling for energy-consuming products imported to Canada or shipped across provincial borders.

The EE Act also grants the federal government the authority to establish programs and activities that promote the efficient use of energy. Moreover, it defines the mechanisms that shall be implemented and used to report the impacts of the Act. Additionally, it instructs the federal government to collect data on energy use. The Act was amended in 2009 to allow the federal government to also prescribe standards for other types of products that affect energy use, such as thermostats.[5]

In 1995, the EE Regulations came into force, thereby establishing not only energy performance levels for four categories of products, but also reporting, testing and labeling procedures. The Regulations apply at the federal level, and in some cases provincial and federal regulations differ or apply to different types of energy-using equipment. Importantly, federal regulations do not supersede provincial regulations.

The Regulations are administered by NRCan and amended frequently to add new products, harmonize federal MEPS with those of other jurisdictions, as well as update testing methodologies and labeling requirements. In June 2017, the new Energy Efficiency Regulations 2016 came into effect, thereby repealing and replacing the 1995

Energy Efficiency Regulations enacted under the Energy Efficiency Act (Amendment 13).

EE in the Building Code

The introduction of EE measures in the building sector was initially assessed and carried out in the late '70s. In 1978, the Measures for Energy Conservation in New Buildings code was published. This prescriptive code was based on the ASHRAE 90 Standard. However, the code had limited success; Quebec was the only province to enforce it, and the CMHC, the only other authority to adopt this code, made compliance mandatory for housing financed under the National Housing Act.

In 1989, several federal and provincial agencies and electric utilities agreed to provide the necessary funding to support research to update the Measures code. In 1997, Canada approved the Model National Energy Code for Buildings (MNECB) which defined minimum thermal performance levels for new buildings based on climate, energy prices and construction costs for commercial and residential buildings. The code was developed by a committee under the CCBFC in close collaboration with energy utilities, provincial and territorial governments, as well as NRCan. Three compliance paths were defined: prescriptive requirements, trade-offs and performance. By 31 March 1997, only the City of Vancouver had formally committed to adopting the MNECB.

In 2007, NRCan and the Building Energy Codes Collaborative (BECC) filed a motion with the CCBFC to update the MNECB, which was approved.

In 2011, a new buildings energy code was published: The National Energy Code of Canada for Buildings (NECB). This code was estimated to be 25% more stringent than the 1997 MNECB. It outlined the minimum EE levels for all new buildings and also offered three compliance paths: prescriptive, trade-offs and performance. The 2011 NECB included requirements for building envelopes, heating systems and equipment, ventilation and air-conditioning, domestic water heating, lighting, as well as for the provision of electrical power systems and motors.[9]

The 2011 NECB was updated in 2015 to include over ninety changes that improve the overall energy performance of buildings. It also updated performance path modeling rules and provided guidance to reflect changes to the prescriptive path as well as more current typical-use building profiles.

Since provinces and territories hold jurisdiction over construction regulations, they are responsible for adopting these buildings codes.

Therefore NECB is a model for provincial and territorial regulations, and it may be amended and/or supplemented by provincial authorities and then published as a provincial code.

The 2011 NECB is currently in force in four provinces (British Columbia, Alberta, Manitoba and Ontario) and two cities (Vancouver and Whitehorse). The 2015 NECB is currently only in force in Nova Scotia.

Minimum Energy Performance Standards (MEPS)

MEPS are an important tool for reducing energy consumption in the residential sector by causing the Canadian market to phase out the most inefficient equipment. MEPS were introduced in 1995 when the 1995 Regulations took effect. MEPS were introduced originally for:

1. Major residential appliances: Electric clothes dryers; clothes washers; dishwashers; refrigerators, freezers and combined refrigerator-freezers; as well as electric and gas ranges.

2. Space conditioning equipment: Room air conditioners; single-package and split-system air conditioners and heat pumps; ground or water-source and internal water-loop heat pumps; and gas furnaces.

3. Water-heating equipment fired by oil, gas and electricity.

4. Lighting equipment: LED lamps, fluorescent lamp ballasts, etc.

5. All types of electric motors.

6. Other energy-using equipment mainly used in the commercial and industrial sectors.

To ensure that MEPS remain effective tools, NRCan regularly amends the Regulations to strengthen the requirements for prescribed products, thus allowing minimum standards to evolve in line with market transformations. Up to now, the Regulations have been updated more than twelve times to include new equipment and modify requirements. As a result, MEPS have been established for more than 40 product categories.

To update the Regulations, NRCan conducts several studies to assess how changes would impact the economy, the environment and society. Once an amendment proposal is defined, it typically undergoes a public consultation process that starts with publication, in the Canada

Gazette (the official newspaper of the Government of Canada), of the Notice of Intent pursuant to a pre-consultation and pre-publication process and ends with the final version being published in the Canada Gazette, Part II.

Three main mechanisms have been developed to ensure compliance with the Regulations. The first is the EE Report which must be sent to NRCan before an energy-using product is imported into Canada or shipped between provinces. With the information contained in the Report, NRCan verifies that the product meets the prescribed EE standard. The CBSA regularly sends information about the imports of regulated products to NRCan which then cross-matches this information with their database to determine compliance. NRCan can instruct customs officials to stop the importation of a product that does not comply with the Regulations.

The second mechanism is the Customs Release Form which must be submitted to the CBSA at the moment of import. It includes information about the imported energy-using product. If the document is incomplete, the customs officer may refuse to allow the product to clear customs.

The third mechanism is the Energy Efficiency Verification Mark. Regulated energy-using products imported into Canada or shipped between provinces must display an EE verification mark from a certification body accredited by the Standards Council of Canada (SCC).

Harmonizing standards between provinces (the governments of Nova Scotia, New Brunswick, Quebec, Ontario and British Columbia regulate energy-using products) and with neighbor states is also a priority. Harmonized standards reduce barriers to trade and avoid confusing customers, thus increasing the effectiveness of standards. Under this framework and as a result of the Canada-United States Regulatory Cooperation Council's Joint Forward Plan, the Canadian and American governments agreed in 2014 to align new standards and update existing ones to reduce unnecessary regulatory differences and facilitate trade.

Natural Resources Canada estimates that, as of January 2016, Canada's regulations align with fewer than 50% of product categories regulated in the United States.[6]

Labeling

In synergy with the introduction of MEPS, Canada also implemented vast labeling programs. Currently, two labeling programs provide information about the energy performance and energy use of products, namely EnerGuide and ENERGY STAR®.

The Canadian label program, EnerGuide, was launched in 1978 as a voluntary program. After the 1995 Regulations came into force, labeling became mandatory for hundreds of home appliances. EnerGuide is a comparative label that permits consumers to compare the energy performance of different appliances of the same category prior to purchase. For example, the EnerGuide label provides information on the total annual energy consumption of refrigerators, but provides an EE ratio for room air conditioners rather than total annual energy consumption. This benefits the consumer.

NRCan also administers the ENERGY STAR® labeling program which is not a comparative label, but rather an endorsement label that allows consumers to identify the most energy efficient products in a given category. This voluntary label was introduced in 2001 and today covers 70 categories of residential and commercial products and identifies the best EE performers (top 15 to 30%).

Furthermore, Canada has implemented EnerGuide initiatives to promote EE in the housing sector. For instance, the EnerGuide Rating System for New Homes is a useful tool for new and renovation projects. It comprises an evaluation tool and an EnerGuide label sticker that displays the results of an EE evaluation.

The EnerGuide Rating System has been updated recently. The scale, originally a 0-to-100 efficiency indicator, has been replaced by an indicator of total annual energy consumption (in gigajoules), which allows for more intuitive comparisons of buildings.

The labeling process starts with a visit from an NRCan-certified energy advisor who performs an EE assessment. During the assessment, detailed information about the home structure and appliances is collected, such as the number and locations of all windows and exterior doors, as well as the size and efficiency of space heating and air-conditioning equipment. This information is then used to assess the energy consumption of the house through a simulation that takes into account standard conditions and occupation levels. The OEE develops and manages the HOT2000 tool used for such simulations.

Once the assessment is complete, the homeowner receives the rating results. If improvements are implemented thereafter, the homeowner may request a follow-up assessment through which a new rating may be issued if appropriate.

The EnerGuide label is also used across Canada to define eligibility for grants and rebates.

In addition to EnerGuide for New Homes, homeowners have access to the ENERGY STAR® for New Homes Program. This certification, launched in 2005, is designed to identify the most energy efficient homes. In 2012, NRCan published the ENERGY STAR® for New Homes Standard - Version 12. Homes built to this standard are on average 20% more energy efficient than homes built to code. Two paths are available to qualify for this label. The first, the performance path, consists of performing an energy evaluation using NRCan software. Under this path, the home may also receive the EnerGuide label. The second, the prescriptive path, consists of choosing among the available building option packages according to location. These packages are designed to ensure a certain level of energy performance.

The R-2000 label is a voluntary standard for new homes that aims to build highly efficient and environmentally friendly houses. R-2000 was developed by NRCan in coordination with key industry stakeholders. Compliance with this label results in newly built homes that are on average 50% more energy efficient than typical new homes. Interestingly, the EnerGuide Rating System is used to assess the efficiency of these homes.[5] There are now more than 10,000 R-2000 homes and close to 900 builders with R-2000 licenses. The R-2000 program is managed by the OEE.

Furthermore, NRCan has implemented a label for new vehicles through the EnerGuide initiative. This label is designed to provide model-specific information on the fuel consumption, annual fuel cost and carbon dioxide rating for new light-duty vehicles available for retail sale in Canada. This information is collected in partnership with ECCC which monitors the emissions of new light-duty vehicles sold in Canada.

Demand-side Management/Demand
Response Program Evaluation

In 1991, the federal government launched the NEUD. This database provides comprehensive information on energy use and GHG emissions for the residential, commercial, institutional, industrial, transportation and agriculture sectors across Canada. To collect the information required, NRCan has, since 1993, sponsored the SHEU which collects the most comprehensive information to date on the energy characteristics of the Canadian housing stock.

The NEUD is a key element for planning and evaluating EE and emission reduction measures in Canada.

Evaluation policies and programs are key components to success-fully implementing policies. The 2016 Treasury Board Secretariat (TBS) Policy on Results requires that all departments measure and evaluate their performance and report to the TBS.

NRCan evaluation activities are carried out by the Strategic Eval-uation Division.[xvi] This Division aims to generate evidence-based assessments of the relevance and effectiveness of NRCan programs and policies and is responsible for proposing evaluation plans to the Evalu-ation Committee.[xvii]

Commonly used methodologies include interviews, document and literature reviews, surveys, case studies, as well as other data and economic analyses. Including the planning and approval processes, large evaluations typically take 12 to 18 months to complete.

FINANCIAL MECHANISMS

Financial mechanisms can be an effective tool to foster EE actions among stakeholders, and the Government of Canada has developed some innovative financial schemes that have contributed to the develop-ment of EE nationwide. The following paragraphs present mechanisms such as the CCA, CIPEC and energy performance contracting (EPC).

Industry

In 1994, the Government of Canada introduced the accelerated CCA for certain EE products as an incentive to encourage businesses to invest in specific energy efficient equipment (Class 43.1). Class 43.1 allows an accelerated CCA rate of 30% per year on a declining balance basis for properties acquired after February 21, 1994. In 2005, Class 43.2 was introduced to allow a CCA rate of 50% per year, also on a declining balance basis.[xviii] CCA allows businesses to write off the capital costs of these assets at a faster rate and thus improves the after-tax rate of

---

[xvi] The Strategic Evaluation Division and the Strategic Planning and Reporting Divi-sion are subunits of the Planning, Delivery and Results Branch.

[xvii] The Departmental Evaluation Committee is an NRCan governance committee chaired by the Deputy Minister.

[xviii] The eligibility criteria for these two CCA classes are generally the same, except that Class 43.2 has a higher efficiency standard for cogeneration systems that use fossil fuels than Class 43.1. (NRCan, 2013)

return on such investments.[10], [11]

In addition, the Income Tax Regulations allow certain expenses incurred during the development and startup phases of energy conservation projects to be fully deducted in the year they are incurred. Examples of this type of costs are certain prefeasibility studies, feasibility studies and environmental assessment expenses.

Under the aforementioned CIPEC, NRCan developed the Process Integration Incentive Program. This program offers a financial incentive to encourage companies to use a process integration study to holistically assess all energy savings opportunities. The program covers up to 50% of the total cost of the study up to a maximum of $40,000.

Commercial and Federal Public Buildings

The FBI is a federal initiative implemented in 1991 aimed at assisting federal departments and agencies to reduce the energy consumption and GHG emissions of their facilities.

The initiative fosters the use of energy performance contracts (EPCs) and thereby allows the public sector to undertake retrofits using private-sector capital funding. Two types of contracts can be used by federal entities, first-out performance contracts and shared savings performance contracts. EPCs have two clear advantages: (1) they preclude the need for federal capital funds; and (2) the risk of achieving energy savings is assumed by energy service companies (ESCOs).

The FBI provides tools, training, model documents, as well as program policy advice and procurement assistance. It also maintains what is called the Qualified Bidders List of qualified ESCOs capable of providing energy management services to federal stakeholders. This list ensures that qualified companies can comply with the needed technical expertise and are financially viable.

Since implementation, the program has supported more than 85 retrofit projects, attracting $364 million in private-sector investments.

Housing

The CMHC Green Home Program provides up to a 25% refund to borrowers who purchase either an energy efficient home or an existing home and then carry out EE improvements. Homeowners can also receive a 15% premium refund for a home built to ENERGY STAR® building standards or a 25% premium refund for a home built to R-2000 standards.

Federal

Adapting Canada's infrastructure to climate change and mitigating the impacts thereof are key aspects of Canada's development strategy. These priorities translated into several initiatives designed to finance the transformation of municipal, provincial and federal infrastructure.

As part of the Investment Canada Plan, the government is investing more than $26 billion over 12 years in green infrastructure. Of this amount, $5 billion is earmarked for green infrastructure projects through the Canada Infrastructure Bank[xix] which became fully operational in 2018.

Municipal

The Green Municipal Fund (GMF) was created in 2000 to support actions to reduce GHG emissions at the municipal level. The GMF provides loans and grants to advise project owners and finance initiatives in six different areas including EE.

A budget of $550 million was initially allocated to the GMF through a series of government decisions from 2000 to 2005. An additional $125 million was announced in Budget 2016 and will be added to the Fund in 2017 or 2018. The GMF provides financing for feasibility studies and pilot projects, or capital costs for projects that reduce GHG emissions. Within the buildings sector, the GMF finances building retrofits and the construction of new buildings. Building retrofit projects have to demonstrate the potential of reducing energy consumption by at least 30% compared to current performance. For new constructions, projects must demonstrate an anticipated reduction in design energy consumption of at least 45% compared to the NECB.

The Government of Canada signed an agreement with the FCM, a non-profit organization, to manage the GMF. The Government of Canada, through NRCan and Environment Canada, participates in the governance of the fund, along with representatives from the public and private sectors, including municipal officials and technical experts.

[xix] The Canada Infrastructure Bank aims to attract private-sector and institutional investments to new revenue-generating infrastructure projects that are in the public interest.

Chapter 2

Alberta

Mr. Jesse ROW, Alberta Energy Efficiency Alliance

ENERGY SECTOR

Alberta's energy consumption has been rapidly increasing since the 2001-2003 period mainly due to the industrial sector which consumed 151% more in 2013-15 than in 1995-97. Total consumption approached 50,000 ktoe annually during the 2013-15 period, which represents a 78% increase when compared to 1995-97 levels. The transportation sector grew slightly less than total consumption during this same period (67%), while commercial, residential, and agriculture sector consumption each grew by less than 20%.

Figure 2-1 illustrates average energy consumption in three-year periods from 1995 to 2015.

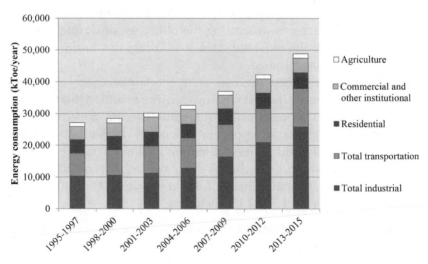

Figure 2-1: Alberta Energy Consumption from 1995 to 2015
Source: Statistics Canada, CANSIM 128-0016

ENERGY EFFICIENCY BACKGROUND

EE initiatives in Alberta have had a long and varied history. As in many jurisdictions, these efforts began decades ago and resulted in both new energy codes for buildings (1981) and the creation of EE programs in the provincial government and utility companies. An assessment of EE activities in Alberta undertaken by the Energy Efficiency Branch of Alberta Energy demonstrated that approximately $8.3 million and 102 person years of resources were dedicated to energy management programs in the province in 1992.[1] Examples of these programs include:

- Old Fridge Roundup by TransAlta Utilities ($1.1 million budget);

- District Conservation Project by Edmonton Public Schools ($900,000 budget);

- High Efficiency Motors Rebate Program by Alberta Power ($500,000 budget);

- Industrial and Commercial Energy Audit Program by the Government of Alberta ($276,000 budget);

- A variety of research, development and demonstration programs.

Unfortunately, the provincial government was undertaking major budget cuts while also restructuring the utility sector during the mid-1990s. As a result, the most substantial EE programs in the province were gradually discontinued.

This trend changed in 2000, with the Government of Alberta establishing Climate Change Central—an organization that, among other initiatives, was responsible for administering government EE programs. Funding varied as provincial budgets cycled between surpluses and cutbacks and political attention to climate change rose and fell (Alberta released climate change strategies in 2002 and 2008). Over its 14-year history, Climate Change Central delivered 23 programs that reduced emissions by 4.5 MT,[2] including:

- Carrying out residential retrofits (with estimated energy savings of $320 million);

- Offering an interest-free loan program for municipalities (loaned a total of $37 million);

- Promoting the use of hybrid vehicles in the taxi industry;

- Supporting fuel efficiency and alternative fuel efforts in the trucking industry;

- Supporting the early adoption of solar energy on municipal building rooftops.

As part of its 2008 climate strategy, the Government of Alberta also established an emissions offset system and the Climate Change and Emissions Management Fund (CCEMF). Both these mechanisms were sources of funding for a handful of EE initiatives in the private sector, but have not been a major source of funding in recent years.

Prior to 2011, ATCO Gas, a natural gas distribution company in the province, delivered modest demand-side management (DSM) programs including education and outreach activities. In 2010, it was proposed that this $1.6 million effort be increased to close to $4 million through the ATCO Gas 2010-2012 General Rate Application. However, in 2011 an important decision was released by the Alberta Utilities Commission (AUC) regarding EE in Alberta. The decision denied cost recovery for the ATCO Gas total DSM budget because DSM activities are not explicitly listed within the legislation governing Alberta's utility system.

In March 2012, the last province-wide EE program for public participation was discontinued by Climate Change Central. Since then, specialty programs have been offered for the municipal and agriculture sectors, indigenous communities, as well as a loan program for seniors (the last two were established in 2016), but no large-scale programming had been offered in the province prior to 2017.

EE programs in Alberta are now changing, however, as the provincial government released its Climate Leadership Plan at the end of 2015. Under this plan, the government indicated it would use a portion of revenues ($645 million over 5 years) from a new carbon levy on consumer fuels (e.g. natural gas, gasoline, and diesel) to fund EE programs. The government has recently established a new agency to administer these programs, Energy Efficiency Alberta, and has launched an initial set of programs throughout the province:

- Residential Instant Savings Program—Offers instant discounts to residential customers at participating retailers on an assortment of energy efficient products including lighting products, smart power bars, timers and water saving devices.

- Residential Online Rebates Program—Offers rebates to residential customers on the purchase of qualifying energy efficient appliances and smart thermostats.

- Residential Home Improvement Rebates Program—Offers rebates to residential customers on qualifying insulation, windows and hot water tanks. Rebates are based on the amount of energy saved and customers must work with a registered contractor to qualify.

- Residential No-Charge Energy Savings Program—Offers in-home advice on EE opportunities and installation of low-cost energy efficient products to households across the province.

- Business, Non-Profit and Institutional (BNI) Energy Savings Program—Offers incentives that encourage organizations to replace outdated and inefficient heating, lighting and process equipment.

- Residential and Commercial Solar Program—Offers rebates to homeowners, businesses and non-profits that install solar photovoltaic (PV) systems in accordance with the Alberta Micro-Generation Regulation.

STAKEHOLDERS

The main stakeholders involved in EE are the Government of Alberta and Energy Efficiency Alberta, a new provincial Crown Corporation responsible for administering EE programs in the province.

Government of Alberta efforts on EE are led by the Alberta Climate Change Office (ACCO). The ACCO is responsible for administering the carbon levy of the province and working with other departments on the distribution of revenues, which includes levy rebates for consumers, funds for emission reduction initiatives, and assistance to communities that are being impacted by the phase out of coal-fired power.

The ACCO was also the lead on setting up Energy Efficiency Alberta; the latter is tasked with developing and delivering programs to:

- Raise awareness among consumers about energy use and its impacts on the economy and environment;

- Design, promote, and deliver EE and conservation programs, as well as develop community energy systems;

- Promote the development of an EE and community energy services industry.

Energy Efficiency Alberta will need to determine, as it establishes itself in the market, how to work with other provincial actors such as energy distributors and retailers, product and service providers, as well as stakeholders.

The Alberta Energy Efficiency Alliance is an industry association whose membership base includes product and service providers, energy distributors and retailers, municipalities, non-profit organizations, and individuals. Its main objective, since its inception in 2007, has been to promote the creation of new EE programs and policies in the province and help build relationships between organizations interested in EE.

Traditional EE program delivery agents in the province are electricity and natural gas distributors and retailers. While these companies have been relatively hands-off with respect to EE in recent years, they have the potential to significantly contribute to future efforts.

Electricity and natural gas systems in Alberta are operated by a combination of regulated and competitive entities. Energy retailers consist of both regulated and competitive retailers for both fuels. Electricity and natural gas transmission and distribution utilities are both regulated, while electricity and natural gas suppliers operate in a competitive market. Currently, none of these organizations plays a formal role in provincial EE programs, although some energy retailers and utilities conduct their own outreach activities, as well as offer EE products and services. These include heating and cooling products, maintenance services, home and business energy reports, general energy saving tips and energy audit services.

The City of Medicine Hat's utility system is one exception to the approach of utilities to EE programs. The local municipal government has historically provided both electricity and natural gas services to residents by supplying fuel mainly from its own natural gas wells. Affectionately nicknamed the Gas City by locals, Medicine Hat's utility system has evolved, to some extent, separately from the rest of the province. This has allowed the local utility to determine its own approach to EE. As a fully integrated utility with a close connection to its customers, it has determined that it is cost effective to create local EE programs and reduce the need to build new energy supply infrastructure. This has led the municipality and utility to work together to run EE information and rebate programs funded through charges on high-use natural gas and electricity bills. Between 2008 and 2015 for example, the HAT Smart program awarded over 14,000 rebates worth more than $4.4 million.[3]

POLICIES AND STRATEGIES

Currently, the most prominent government strategy for EE is the Alberta 2015 Climate Leadership Plan. This plan outlines approaches for reducing GHG emissions and includes the creation of the aforementioned Energy Efficiency Alberta. As Energy Efficiency Alberta establishes itself in the market, details on measurement and verification processes will emerge.

The Climate Leadership Plan also includes provincial policies on a carbon levy for consumer fuels such as natural gas, gasoline, diesel, as well as carbon regulations for large final emitters such as large oil sands facilities and power plants. The subsequent section outlines the regulatory frameworks for large final emitters, from both the 2008 and 2015 climate plans.

For its own facilities, the Government of Alberta has been using BOMA BEST since 2005 to help ensure government buildings are managed in an energy efficient manner. In 2006, the LEED Silver requirement was added for all new buildings.[4] The Province also has a regulation enabling school boards to borrow funds for EE upgrades when loans are guaranteed by EPCs and costs are recovered from generated savings within 20 years or less.

LEGAL AND REGULATORY FRAMEWORK

In its 2008 Climate Change Plan, the Government of Alberta proposed both a provincial Energy Efficiency Act and an increase in EE requirements in the building code.[5] While the Act did not move forward, the government decided to adopt national energy codes for buildings rather than establishing its own. The 2011 National Energy Code for Buildings and the 2012 National Building Code which includes EE provisions for houses came into full force in November 2016.

Other EE regulations include provisions affecting large final emitters in the province. Since 2007, Alberta has had a Specified Gas Emitter Regulation (SGER) in place. This regulation requires large emitters (greater than 100,000 tons of CO_2e per year) to reduce their emissions compared to their historical emissions by one of three approaches:

• Reducing their own emissions;
• Purchasing credits or offsets from others;
• Paying into the CCEMF.

While the SGER provides incentives for facilities to reduce emissions through these three mechanisms, most emitters choose to comply through payments to the CCEMF.[6] As previously stated, several industrial EE projects were funded through this fund as part of the activities of the Climate Change and Emissions Management Corporation (CCEMC) early in its mandate. However, since then funding for additional EE projects has been limited because political decisions were made to direct most of the funds into the development of new oil sands technologies. A small number of EE projects have also generated and sold offsets during the early years of the SGER, but transaction costs associated with certifying offset projects limit this mechanism from being used broadly.

The government is now making changes to regulations applicable to large emitters. Most notably, emission targets will be determined by comparing a facility's current emissions to an industry-specific baseline, as opposed to its historical emission levels.

FINANCIAL MECHANISMS

At the time of writing, there are relatively few EE specific financial mechanisms in Alberta.

Existing incentive programs are those offered by the Municipal Climate Change Action Centre, Alberta Agriculture, Alberta Indigenous Relations, as well as Alberta Seniors and Housing. Other incentive programs recently launched with the creation of Energy Efficiency

Alberta include rebates for a number of lighting and heating products for the residential and commercial sectors, as well as rebates for residential insulation, tankless water heaters and window upgrades.

Additional financing programs are expected to be launched by Energy Efficiency Alberta, which could include new approaches to innovative financing. For example, Alberta Municipal Affairs has attempted to add Clean Energy Loans to the Big City Charters being developed for Calgary and Edmonton,[7] although the final form of innovative financing programs is yet to be determined.

Chapter 3

British Columbia

Mr. Andrew PAPE-SALMON and Mr. Tom BERKHOUT
Government of British Columbia

ENERGY EFFICIENCY BACKGROUND

In 2015, the three major sources of energy consumed in British Columbia (BC) were electricity (192 PJ), natural gas (237 PJ), and petroleum-based products (393 PJ).[1] More than 90% of the electricity generated by BC-based utilities came from hydroelectric sources, less than 5% from fossil fuels, and the remaining 5% from other renewable sources of energy.[2] BC Hydro provides electricity services to over 95% of the province's electricity customers. FortisBC Inc., hereafter referred to as FortisBC (electric), services most of the remaining south central parts of the province. Five municipal utilities also provide distribution services, one of which generates electricity.

Two utilities distribute natural gas in BC: FortisBC Energy Inc., hereafter referred to as FortisBC (gas), which services over 96% of the province's natural gas customers, and Pacific Northern Gas which services parts of northern BC.

Policies and measures to encourage EE have been actively pursued in the province since the late 1980s. To date, almost all efforts have focused on the industrial, agricultural and building sectors. Limited efforts in the transportation sector have focused on reducing GHG emissions.

EE is actively pursued by major electric and natural gas utilities, the provincial government and certain local governments. EE is an important strategy for reducing energy bills and GHG emissions, expanding economic opportunities and competitiveness, creating jobs, as well as enhancing community resiliency.

The business drivers for DSM[i] vary among gas and electric

---

[i] DSM in British Columbia is pursued under the auspices of the Demand-Side Measures Regulation.

utilities. DSM is generally considered an energy resource by electric utilities, provided that incremental energy supply costs exceed the cost of conserved energy, and/or capacity increases can be delayed. Gas DSM is also considered a resource, but gas utilities are primarily delivery agents and pass commodity costs on to the consumer. Other drivers for gas DSM include GHG reduction benefits. For both types of utilities, a primary driver of conservation activities is customer service to help manage energy bills because of rate increases.

A Brief History

Although mentions of EE and conservation are found in provincial policy documents that date back to the early 1980s, BC's pursuit of energy savings began in earnest in 1988 when BC Hydro launched its Power Smart program. Shortly thereafter, the program was fully institutionalized into the fabric of the provincial energy plan when BC Hydro put DSM on an equal footing with supply-side resources in the utility's first ever integrated resource plan (IRP). At the time West Kootenay Power, now FortisBC (electric), also launched a DSM program and the BC Energy Efficiency Act was proclaimed.

BC Hydro has maintained an integrated resource approach to determine the size of its DSM portfolio, with the exception of a period during the late 1990s and early 2000s when DSM was temporarily significantly scaled back in response to policy deregulation.

In the case of FortisBC (gas), the pathway to providing natural gas DSM programs was more gradual. The utility launched its DSM program in 1997.

Running in parallel to these utility programs have been a number of provincially led initiatives to improve EE, mostly in partnership with energy utilities. The BC Energy Efficiency Act ([RSBC 1996] CHAPTER 114) sets out energy performance standards for designated appliances and equipment. The provincial government is also responsible for energy policy that affects utility DSM programs, transportation emission reductions, building energy codes, as well as local government enforcement of codes and advancement of EE. Through the Innovative Clean Energy (ICE) Fund, the Province is currently involved in a number of market transformation initiatives to promote clean energy vehicles, high performance buildings and improved industrial energy management practices.

A Market Transformation Approach

The above review of BC's EE history demonstrates an integrated approach, including utility-led resource acquisition and government-led policy, to enable market transformation (MT). MT is "a policy objective to encourage or induce social, technological and economic change toward greater energy efficiency."[3] The idea is to permanently transform the entire marketplace by reducing market barriers to the development and diffusion of more efficient technologies, processes and practices. Furthermore, MT implies an eventual sunset of financial investments by governments and utilities following the introduction of regulated codes and standards.

Early experience in BC with transforming the marketplace for discrete technologies such as high-efficiency motors, furnaces and light bulbs demonstrates the value of an integrated approach to both DSM programs and eventual codes and standards. An MT approach generates long-term savings that are both substantial and more cost-effective than approaches with a shorter timeframe and narrower scope of measures. Today, the Province, utilities, communities and a host of other actors are taking the lessons learned from earlier MT initiatives to support the target of a net-zero energy ready[ii] new construction standard by 2032.

In most cases, MT and low-cost EE measures are similar: research and development, pilot projects, training, incentives, voluntary leadership, regulation and enforcement. They differ on two fronts. MT involves the intentional and ongoing coordination of the different actors and the measures being implemented, while low-cost EE measures involve the explicit eventual withdrawal of subsidies commonplace in DSM.

STAKEHOLDERS

The EE framework that has evolved over the last 25 years in British Columbia has been significantly influenced by four major types of stakeholders:

[ii] Defined as maximizing economically optimal energy efficiency by considering long-term avoided energy costs with the capacity to meet the resulting consumption demand through onsite or community-based clean energy resources.

1.   Electricity and natural gas utilities
2.   Provincial government
3.   Municipalities
4.   Federal government

Electricity and natural gas utilities in BC include: BC Hydro and FortisBC, Pacific Northern Gas, and municipal utilities. The latter do not provide DSM directly. Utilities have been advancing EE initiatives to achieve three goals, namely:

•   Resource acquisition—EE and load management in lieu of new energy supplies and, in some cases, capacity increases;

•   Customer services—information and incentives to help consumers decrease energy bills and increase competitive edge; and

•   Market transformation—enabling voluntary leadership and codes and standards.

The BC Ministry of Energy and Mines (MEM) advances energy planning, coordination among multiple players, the utility regulatory framework for DSM, and equipment EE standards. The Ministry of Municipal Affairs and Housing fosters the adoption of building energy codes in legislation while delegating information and capacity building measures through partners such as BC Housing or the Building Officials Association of BC.

Municipalities in BC advance EE by enforcing building codes or supporting separate climate leadership initiatives for buildings, transportation and industry linked to the provincially led Climate Action Charter. Municipalities also promote EE in their own operations and employ community energy managers and specialists whose salaries are co-funded by utilities.

The federal government, in particular NRCan, has a direct impact on EE in BC through equipment standards under the Energy Efficiency Act (S.C. 1992, c. 36) and the EnerGuide and ENERGY STAR® labels. Furthermore, the Natural Resource Council of Canada (NRC) develops the model national energy codes for buildings.

Institutions in Charge of Strategy

The four abovementioned major stakeholder groups and additional supporting stakeholders partner extensively to promote MT toward optimal EE levels in the economy. Furthermore, they collectively lead different components of the energy strategy in BC, influenced by NRCan on a broad range of issues and the NRC for building codes. Furthermore, MEM reviews regulatory proceedings of the California Energy Commission for product standards. These organizations work together to implement transformative legislation and regulatory initiatives that have positioned BC as a leader in North America.

Institutions in Charge of EE Programs

BC EE programs are almost exclusively delivered by vertically integrated utilities. These are all regulated by the BC Utilities Commission, although oversight of some components of BC Hydro programs is provided by the provincial government. Energy utilities are the primary organizations that deliver DSM programs and cover the full spectrum of programs and initiatives in place across North America. In recent years, there has been an increased emphasis on coordinated and jointly funded programs that target the same demographic to phase out the use of different fuels.

Between 2008 and 2012, the provincial government ran a major program for existing detached houses called the LiveSmart BC: Efficiency Incentive Program in partnership with utilities, which at one time or another included co-funding from the federal ecoEnergy program.

POLICIES AND STRATEGIES

Since 2007, the primary drivers of BC EE policy have been the provincial climate change and clean electricity agendas. Although utility resource savings programs still play central roles, the context for utility resource planning is very much framed within broader climate objectives.

Provincial Energy and Climate Policy Framework

In 2007, the Government of British Columbia announced its commitment to reduce its 2007-level GHG emissions by 33% by 2020 and 80% by 2050. In that same year, the province released the BC Energy Plan—A Vision for Clean Energy Leadership. The Plan included a number of measures that set the stage for expanding efforts to pursue

EE and conservation in BC, including the following set of commitments that impacted utility DSM, among others:

1. Establishing a target for BC Hydro to acquire 50% of its incremental resources (GWh) through conservation efforts by 2020. This target was subsequently changed in the 2010 Clean Energy Act to 66% of incremental load growth and has been consistently exceeded by BC Hydro. Also, FortisBC (electric) voluntarily adopted this target and has subsequently increased it to 80% of GWh growth by 2023;

2. Ensuring a coordinated approach to conservation and efficiency—for example, BC Hydro included government actions in their resource and DSM plans;

3. Encouraging both electric and natural gas utilities to pursue cost-effective DSM opportunities, supported by amendments to the Utilities Commission Act to prioritize DSM and enable cost recovery, along with the development of the DSM Regulation; and

4. Developing rate structures that encourage EE and conservation.

The first measure further solidified the importance of DSM as a strategic energy resource for BC Hydro. The second policy objective was led by the MEM and was embraced by the two major utilities and officially endorsed by the BC Utilities Commission, eventually leading to multi-fuel DSM programs. The third policy objective broadened the type of DSM that utilities could pursue (i.e. within the bounds of cost effectiveness as defined in the DSM Regulation), which signaled there was a desire to see an expansion of both gas and electric DSM programs. Finally, the fourth commitment provided utilities with a new tool to foster energy conservation, namely a rate-based signal for electric utilities to encourage industrial, commercial and residential customers to reduce their total energy consumption. However on April 1, 2017, BC Hydro removed the conservation rate for commercial customers, but retained it for residential and industrial. The commercial conservation rate was removed in a Rate Design Application due to poor performance in achieving conservation, largely due to its complexity.

A second set of policy commitments made in the 2007 Energy Plan targeted EE specifically in the built environment:

1.   Implement EE standards for buildings (by 2010) by introducing
      regulated standards in the BC Building Code and advancing
      ongoing efforts to develop energy efficient products through reg-
      ulated standards;

2.   Pursue public-sector leadership of environmental design for new
      buildings; and

3.   Encourage increased community-level EE and GHG reductions
      for both its own operations and private buildings.

These three built environment measures largely constituted a
progression of the MT efforts outlined in the province's 2005 Energy
Efficient Buildings Strategy, which were expanded in the subsequent
2008 Strategy.

On the industrial front, the 2007 Energy Plan included a commitment to developing an industrial EE program to address specific
challenges faced by the BC industrial sector. This initiative was essentially delegated to energy utilities, including expanding BC Hydro
and FortisBC (electric) programming to all industrial scales and new
FortisBC (gas) programs.

Many of the energy and climate policy directions established in
2007 and 2008 were subsequently embedded within different pieces of
legislation and supported by numerous programs. In 2010, the Clean
Energy Act added GHG emission reductions to the list of provincial
energy policy objectives, including a 33% reduction below 2007 levels
by 2020.

In the early 2010s, market and legislative changes led to a supply surplus and reduced the need for DSM. Furthermore, in 2011 a
provincial audit, Review of BC Hydro,[4] examined all BC Hydro
expenditures and the impact of DSM measures on rates, as well as
other financial indicators. Despite these factors, the BC Hydro 2013 Integrated Resource Plan (IRP) recommended a moderate DSM portfolio
consistent with the 2008 Long-term Acquisition Plan, albeit expanding
DSM portfolios was considered.

The most recent set of major policy objectives impacting EE in BC
was released as part of the 2016 Climate Leadership Plan, namely:

•   Encouraging the development of high-efficiency buildings by
      setting a target for a net-zero energy ready building standard by
      2032, establishing an Energy Step Code for new buildings, and

providing a design incentive for high-performance buildings;

- Updating equipment standards for space and water heating equipment by 2020 and 2025 respectively;

- Advancing efficient electrification;

- Updating the Climate Action Charter for Communities; and

- Enabling FortisBC (gas) to expand DSM program incentives by at least 100%.

Utility Resource Planning Policy Framework

For the most part, electricity and natural gas in BC have been regulated by the British Columbia Utilities Commission (BCUC) since 1980.

Legislation and BCUC policies require utilities to file a long-term (i.e. 20 years) resource plan (LTRP) and associated short-term (i.e. up to 5 years) DSM plan that includes, among other things, objectives for DSM investment (in $M) and avoided consumption (in GWh or TJ) by sector and measure type. Since 2010, the BC Hydro IRP has been exempt from BCUC oversight, a plan similar to an LTRP.

Utility DSM expenditures require approval from the BCUC before rates can be increased to recover costs. The DSM Regulation establishes rules that the BCUC must follow when assessing the adequacy and cost effectiveness of proposed DSM expenditures.

One major difference in the regulatory context between natural gas and electric utilities is the application of what is called the Modified Total Resource Cost (MTRC) test. The MTRC is based on the Total Resource Cost (TRC) which is one of the standard DSM cost-benefit calculations used across North America to determine if a measure is cost effective (i.e. costs associated with a DSM program are less than benefits such as the avoided cost of energy saved). In the case of the MTRC, up to 40% of natural gas utility DSM savings is valued as the avoided cost of clean electricity rather than the avoided cost of natural gas which is historically lower. The intention of the MTRC is to assign explicit value to GHG emission reduction whenever a unit of natural gas is saved and assist with enabling cost-effective DSM in periods of low-cost natural gas. As a result, a significant portion of natural gas utility DSM portfolios in BC are justified on the grounds of GHG reductions. The MTRC also includes a 15% adder for non-energy benefits. The DSM portfolios of electric utilities that generate nearly 100% of their electricity from renewable sources are justified primarily because of their associated supply-side savings.

Once approved, these long-term planning documents and their associated expenditure schedules provide considerable information about the strategic scope of each utility DSM program. In 2013, BC Hydro's cumulative long-term savings target for its DSM program was 6,306 GWh/year from 2013 to 2021. This has since been adjusted in the 2016 Revenue Requirements Application which includes a different base year. A number of different strategies are used by the utility to achieve these savings for each of its major rate classes, specifically: codes and standards support, rates and programs. BC Hydro is also currently pursuing pilot studies on load curtailment and demand response and is expected to implement a low carbon electrification program as per the BC Climate Leadership Plan.

Until 2008 FortisBC (gas) operated a modest DSM program, valued at approximately $4 million per year, which focused mainly on condensing gas furnaces. In 2008, a landmark BCUC application established a comprehensive program across all sectors that has since grown to over $30 million per year. The most recent FortisBC (gas) LTRP[5] projected cumulative savings of 23 PJ from 2014 to 2018.

Table 3-1 outlines how these savings are distributed across different program components as well as the distribution of expenditures to achieve savings. In response to the 2016 BC Climate Leadership Plan, FortisBC (gas) DSM conservation expenditures are expected to increase substantially.

The 2016 FortisBC (electric) LTRP indicates a trend toward greater reliance on DSM over the next ten years to meet long-range resource needs, gradually increasing from expected annual savings and expenditures of 25.7 GWh/year and $7.6 million respectively in 2017 to 32 GWh/year and $10.9 million by 2023.[6] This plan is expected to account for 77% of projected load growth over this period—well over the 66% target that was set for BC Hydro in the Clean Energy Act. Moreover, this plan is supported by amendments to the DSM Regulation.

Table 3-1 compares each utility's anticipated cumulative savings and expenditures and provides some helpful insight into their strategic similarities and differences.

BC Hydro collaborates extensively with all levels of government

Table 3-1: Comparison of Cumulative DSM Savings and Expenditures

Program Area	% of Cumulative Savings			% of Expenditures		
Utility	BC Hydro[7]	Fortis Gas[8]	Fortis[9] Electric	BC Hydro	Fortis Gas	Fortis Electric
Years of Plan	2014-21	2014-18	2012-16	2014-21	2014-18	2012-16
Codes and Standards	37%			2.1%		
Rate Structures	18%			1.9%		
Residential Programs	5%	29%	48%	14%	39%	40%
Commercial Programs	12%	46%	46%	25%	31%	35%
Industrial Programs	27%	21%	6%	41%	7%	4%
Innovative Technologies		4%			3%	
Enabling Activities	N/A	N/A	N/A	16.1%	20%	22%

and organizations that develop standards to support the establishment of EE regulations. Industrial savings, meanwhile, are largely targeted through incentive programs.

Municipal Government Policy Framework

BC municipal governments represent another group of actors with the authority and resources to influence a broad range of EE and conservation measures. This authority is granted under the Local Government Act and specified within the *Community Charter* for all local governments, except for the City of Vancouver which has separate authority granted under the Vancouver Charter.

For many regional local governments, the motivation to reduce energy use both within their own operations and across their communities is derived from targets included in official community policies. Since 2007, 98% of local governments in BC have signed onto the province's Climate Action Charter. Under the Charter, local governments commit to working toward carbon neutral operations, measuring community-wide emissions, and creating complete, compact, and energy efficient communities.

The City of Vancouver has more authority than any other local government in the province to pursue EE, including establishing its own energy requirements for new and existing buildings. Vancouver has enacted a number of policies for reducing energy consumption and carbon emissions and supported these policies through a range of measures including stringent energy performance building code requirements. Although Vancouver's policies are led by carbon reduction targets, EE is recognized as a foundational strategy for achieving these.

LEGAL AND REGULATORY FRAMEWORK

Provincial and Municipal Framework

The following laws comprise direct stipulations for EE and conservation:

- Energy Efficiency Act ([RSBC 1996] CHAPTER 114) and the Energy Efficiency Standards Regulation;
- Utilities Commission Act ([RSBC 1996] CHAPTER 473) and the DSM Regulation;

- Vancouver Charter ([SBC 1953] CHAPTER 55);
- Community Charter ([SBC 2003] CHAPTER 26);
- Clean Energy Act ([SBC 2010] CHAPTER 22); and
- Building Act ([SBC 2015] CHAPTER 2).

The Energy Efficiency Act (EEA) was enacted in 1990. It allows the province to regulate products sold, manufactured or leased in BC that control or affect the use of energy. It also provides authority for the Province to set specific testing, labeling and reporting requirements for these products. The Act complements the federal Energy Efficiency Act (S.C. 1992, c. 36) which is restricted to regulating the efficiency of selected products that move across national or provincial borders.

The Utilities Commission Act (UCA) was enacted in 1980 and was most recently amended in 2015. It stipulates the administrative framework and responsibilities of the BCUC. Sections 44.1 (Long-term Resource and Conservation Planning) and 44.2 (Expenditure Schedule) of the UCA are the most directly relevant to EE in the province.

Section 44.1 requires utilities (other than BC Hydro) to submit to the BCUC a long-term resource plan (LTRP) that includes, among other things, an estimate of the energy demand the utility expects to serve, a plan on how it intends to reduce the estimated demand, an estimate of the remaining demand and how it intends to meet it. This sequence of actions is important because it requires utilities to first consider and demonstrate how it plans to fulfill anticipated future demand with demand-side measures. Only after this is accomplished are other resources considered. Once submitted to the BCUC, the LTRPs are typically subject to public hearings. The commission then rules on the appropriateness of LTRPs, as per the UCA, taking into account the evidence provided by the utility and interveners at public hearings.

One exception to the regulatory process outlined above is that as of 2010 BC Hydro is required to submit its own Integrated Resource Plan (an LTRP) to the MEM, not the BCUC, for approval. It is then the responsibility of the BCUC to rule on the appropriateness of any expenditure applications (other than expenditures exempted from BCUC review by the Province).

Section 44.2 of the UCA addresses the process for obtaining BCUC permission to proceed with a schedule of planned utility capital expenditures, including demand-side measures, and to recover those expenditures through rates. Utility expenditure plans, which are submitted to the BCUC separately from LTRPs, provide greater levels of granularity on DSM program types and target sectors. All BCUC applications are scrutinized by intervener groups and BCUC staff prior to a decision rendered by Commissioners.

The BCUC is guided by the DSM Regulation which sets out select rules that the commission must follow when assessing the adequacy and cost effectiveness of proposed DSM plans. The Regulation was enacted in the fall of 2008.[10] Since then, it has been amended three times: December 2011, July 2014 and April 2017. Some key elements of the regulation include:

- A requirement for utilities to operate programs for low-income households, rental accommodations, schools, codes and standards support, as well as for governments to adopt Energy Step Code;

- Portfolio-level evaluation of education programs, EE training, community engagement, technology innovation programs, and effective public awareness programs;

- The ability for utilities to attribute a portion of savings from a regulated standard to a utility program that facilitates or advances the introduction of that standard;

- On average at least 1% of DSM expenditures, or $2 million per year, must be dedicated to codes and standards support;

- A requirement for the Total Resource Cost test to be used to determine cost effectiveness and a stipulation for how the avoided cost of supply is determined;

- A requirement that FortisBC (electric) use its long-run marginal cost of clean BC electricity (rather than the short-term spot market price) for the avoided cost of energy;

- Permission for the use of an MTRC for a portion of the portfolio. The MTRC permits the avoided cost of natural gas savings to be priced at BC Hydro's long-run marginal cost of electricity generated from clean or renewable sources;

- Inclusion of estimates of the non-energy benefits (NEB) associated with each planned DSM measure.

Finally, the UCA requires utilities to report on the effectiveness of DSM portfolios through an annual report to the BCUC. As such, all utilities in BC with DSM programs have evaluation as well as measurement and verification initiatives in place.

The Vancouver Charter and Community Charter provide powers to municipalities and regional districts to advance EE and GHG management via their community plans, regional growth strategies and building construction regulations.

The 2010 Clean Energy Act (CEA) is a broad-reaching piece of legislation comprising 16 energy objectives that include a target for BC Hydro to acquire at least 66% of its incremental load growth through conservation by 2020.

For new construction and major retrofits, the Building Act is arguably the most important piece of legislation for advancing EE; it includes the BC Building Code and any EE requirements that fall under the Code. The Community Charter and the Vancouver Charter afford local governments the ability to administer and enforce provincial building requirements. All local governments, with the exception of Vancouver, are limited to referencing the building requirements included in the provincial Code. Vancouver is able to set its own building requirements under the Vancouver Building Bylaw.

Buildings and Equipment

At least three BC-specific regulations or bylaws set MEPS for products, equipment and buildings in the province:

1. Energy Efficiency Standards Regulation (EESR);
2. BC Building Code;
3. Vancouver Building Bylaw.

The EESR lists the testing and performance standards for more than 40 products regulated under the Energy Efficiency Act. Major product categories covered by the EESR are:

*   Manufactured fenestration products;
*   Household appliances;
*   HVAC products;
*   Water heaters;
*   Lighting products;
*   Electric motors.

Since it was first adopted in 1990, five amendments have been made to the EESR. Public consultation on a proposed sixth amendment took place in the fall of 2016. The 2016 Climate Leadership Plan also laid out policy targets for further updates to the performance standards applicable to space and water heating equipment in 2020 and 2025, respectively.

The EESR is unique in that it applies to products manufactured or sold in BC. However, its authority does not extend to building sites, except in the case where a regulated product is assembled onsite. In all other cases, the minimum energy performance of a product used in the construction of a building is regulated by the standards stipulated in either the BC Building Code or the Vancouver Building Bylaw. The scope of these building standards, though, is more limited than the EESR because they do not extend to household appliances, plug loads and other equipment installed post-construction.

As with most provinces, the BC Building Code is largely adopted from the National Building Codes established by the NRC. For energy performance, the BC Building Code 2012 references ASHRAE Standard 90.1 - 2010 and the NECB 2011 for large, complex buildings (Part 3 buildings). For housing and small buildings (Part 9 buildings), the BC Building Code references the 2012 revisions to the National Building Code with some exceptions for BC.

The BC Building Code was amended in April 2017 to include the BC Energy Step Code. The BC Energy Step Code is a voluntary roadmap that establishes five progressive performance targets (i.e. steps) from the current EE requirements in the BC Building Code for net-zero energy ready buildings, which is aligned with the above-mentioned goal for 2032. The BC Energy Step Code is a voluntary tool local governments across BC can use to encourage or require the construction of more energy efficient buildings in their communities in a consistent and predictable manner. It takes a performance-based approach with energy modeling and whole-building airtightness testing, rather than the traditional prescriptive approach. It identifies EE targets that must be met and allows the designer/builder decide how to meet these.[11]

The Vancouver Building Bylaw includes additional requirements beyond the BC Building Code for new construction and major retrofits, such as: higher EE for walls, roofs, windows and skylights; energy-efficient hot water tanks, boilers and furnaces; as well as improved air-tightness in single and multi-family houses. More recently, the City of Vancouver announced the intention to use a phased approach that will eventually require all new buildings built in the city to be zero emissions by 2030. To achieve this, buildings will need to be built to a high-efficient zero-emission standard and/or be heated by a low-carbon energy source such as clean electricity, renewable natural gas or a district energy system that is increasingly fueled by a renewable energy supply.

Greenhouse Gas (GHG)

Several laws and regulations in BC either put a price on GHG emissions or require some level of public disclosure about emission levels. Although these laws and regulations do not directly address EE, regulations often serve as a cost-effective strategy to reduce exposure to carbon-related costs and risks. Likely the best-known law that falls within this category is the Carbon Tax Act ([SBC 2008] Chapter 40) and the accompanying Carbon Tax Regulation. The Act places a price of $30 per ton of CO_2e on nearly all GHG emissions in the province.

The Greenhouse Gas Reduction Targets Act ([SBC 2007] Chapter 42) intended all provincial public-sector operations to be carbon neutral by 2010. To be in compliance with the Act, a public-sector organization must measure its emissions, demonstrate efforts to reduce these through conservation measures, purchase provincially approved offsets for any remaining emissions, and report annually on its progress. Although the Act and its accompanying Carbon Neutral Government Regulation do not stipulate a specific level of energy performance for various public-sector operations, the annual reporting does reveal that these requirements have successfully spurred a number of significant EE improvements.

Another important piece of climate legislation affecting energy use is the Local Government (Green Communities) Statutes Act. Enacted in 2008, it requires local governments to include GHG emission targets, policies and actions in their Official Community Plans and Regional Growth Strategies.

FINANCIAL MECHANISMS

Many DSM programs include financial mechanisms such as incentives and rebates designed to overcome affordability barriers, along with other elements such as education and capacity building to address other barriers. As programs transform the market and enable codes and standards to come into effect, financial mechanisms are reduced or eliminated and programs shift focus to new opportunities to enable the next level of MT.

Incentive Programs

Incentive programs broadly cover the utilities' customer bases. These typically include programs for the following sectors: residential, small and medium general service (commercial, institutional, multi-unit residential), large general service (industrial), and transmission voltage (large industrials with their own substation).

BC Hydro Power Smart

BC Hydro has fully implemented three major cycles of its Power Smart DSM program since 1989. Tables 3-2 and 3-3 further summarize the savings and expenditures of the program by actual (2013-2016) and forecasted (2016-2019) savings.

FortisBC Energy Efficiency Conservation Programs

Recently, FortisBC's gas and electric utilities merged their DSM programs under one banner—FortisBC Conservation and Energy Management. Tables 3-4 and 3-5 summarize the annual savings and expenditures for FortisBC gas (actuals for 2014-16 and planned for 2017 and 2018).

Local Government Development Incentives and the Energy Step Code

Provincial legislation established requirements for GHG reduction targets that include EE. These are frequently achieved through building development approvals. Developments that are within the current zoning bylaw for a city or neighborhood are required to meet the BC Building Code EE standards. However, if rezoning is required for a development, the City can negotiate a higher energy standard in alignment with the Energy Step Code that took effect on April 7, 2017. In fact, EE municipal bylaws that were misaligned with the Energy Step Code ceased to have legal force on December 15, 2017, under the Building Act. Moreover, a number of municipal planning incentives are readily available to developers, such as density bonuses.

Table 3-2: BC Hydro Incremental Annual Electricity Savings (GWh/yr)

Sector	2013-14[12]	2014-15[13]	2015-16[14]	2016-17	2017-18	2018-19
Residential	49	47	46	37	26	28
Commercial	97	83	87	127	51	44
Industrial	122	181	237	131	299	82
Rates	272	(11)	143	23	26	8
Codes and Standards	145	144	557	282	257	314
TOTAL Energy	686	444	1,069	599	659	477
MW Capacity[iii]			130	146	97	100

Table 3-3: BC Hydro Expenditures ($ millions)

Sector	2013-14[15]	2014-15[16]	2015-16[17]	2016-17	2017-18	2018-19
Residential	$17.6	$14.2	$16.0	$13.1	$11.8	$13.0
Commercial	$42.6	$36.7	$33.6	$43.9	$29.9	$25.7
Industrial	$36.1	$45.7	$64.8	$26.7	$84.6	$27.4
Rates	$1.0	$1.2	$1.3	$1.2	$1.0	$1.2
Codes and Standards	$1.6	$3.0	$4.7	$4.7	$4.8	$4.9
Capacity			$8.6	$10.0	$14.2	$14.4
Supporting Initiatives	$21.3	$19.1	$16.1	$14.0	$14.2	$14.2
TOTAL	$120.3	$124.8	$145.2	$113.7	$160.6	$100.7

[iii] Includes capacity benefits of all programs, not just load displacement.

Table 3-4: FortisBC Natural Gas DSM Savings (Annual TJ/yr)

Sector	2014[18]	2015[19]	2016[20]	2017[21]	2018
Residential	94.1	121.4	121.9	136.7	157.9
Commercial	254.9	270.9	255.4	237.7	183.3
Industrial	19.7	16.6	18.3	190.3	189.5
Low Income	24.9	24.1	36.9	27.8	28.2
Conservation Education and Outreach	None attributed				
Innovative Technologies	None attributed	1.6	6.3	5.3	29.5
TOTAL	393.6	434.6	438.8	597.8	588.3

Table 3-5: FortisBC Gas DSM Expenditures ($ millions)

Sector	2014	2015	2016	2017	2018
Residential	$10.9	$12.7	$12.5	$10.7	$11.4
Commercial	$9.4	$10.7	$10.6	$10.4	$10.1
Industrial	$0.7	$1.0	$1.0	$3.0	$3.0
Low Income	$0.9	$1.6	$2.3	$3.2	$3.5
Conservation Education and Outreach	$2.7	$2.8	$2.4	$2.4	$2.4
Innovative Technologies	$0.5	$0.6	$0.8	$1.2	$1.2
Enabling and Portfolio Level	$2.4	$2.4	$2.5	$4.4	$4.4
TOTAL	$27.6	$31.9	$32.2	$35.4	$35.9

Public vs. Private EE Investment and
Leveraging from Incentive Programs

Under the BCUC cost-effectiveness evaluation framework, utilities have estimated the extent of private investment leveraged by incentive programs. For example, the FortisBC Efficient Boiler Program has a utility incentive averaging $13,111 for existing boiler replacements on a total investment of $21,777, thus leveraging a 66% investment by the owner. Also, the Continuous Optimization Program leveraged $2.80 of private funds for every $1 of utility investment in 2016.

Chapter 4

Manitoba

Mr. Dany ROBIDOUX, Eco-West Canada

ENERGY SECTOR

Annual energy consumption in Manitoba was 6,690 ktoe for the 2013-15 period, a 10% increase compared to 1995-97 levels. This rise was mainly due to a 65% energy consumption increase in the industrial sector over the 20-year period. The industrial sector accounted for 21% of the energy mix in 2013-15. For the 20-year period, the second-most energy intensive sector was residential which increased by 12% and represented approximately 19% of total consumption. On the other hand, the commercial sector decreased yearly consumption by 14% over the 1995-2012 period, but then increased in 2013-15 to 2% above 1995-97 levels, which represented 17% of the energy mix for the three-year period. The transportation sector remained the most energy intensive with 33% of Manitoba's total consumption in 2013-15 despite a 4% decrease over the 20-year period. Agriculture represented 8% of the mix, and public administration represented 2%, each decreasing by 2% and 13% respectively between 1995 and 2015.

Figure 4-1 illustrates the average annual energy consumption in three-year periods from 1995 to 2015.

ENERGY EFFICIENCY BACKGROUND

Manitoba Hydro Created in 1961

In 1961, Manitoba Hydro was created as the result of a merger between the Manitoba Power Commission and Manitoba Hydro Electric Board. The town of Emerson wholesale account was among the first to be transferred to Manitoba Hydro. In 1973, Manitoba Hydro purchased the last private mine-owned utility systems in Flin Flon. Also during the early 1970s, tie power lines were installed between Manitoba Hydro

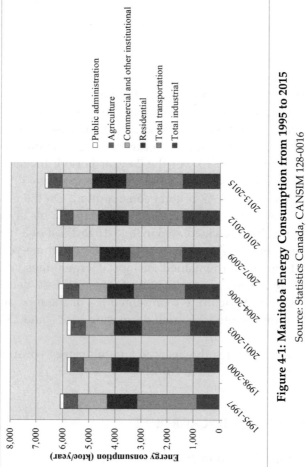

Figure 4-1: Manitoba Energy Consumption from 1995 to 2015

Source: Statistics Canada, CANSIM 128-0016

infrastructure and that of utilities in Ontario, Saskatchewan and Minnesota.[1]

Since the early 1970s, Manitoba Hydro has been the sole commercial provider of electrical power in the province, with the exception of Winnipeg. Until 2002 Winnipeg Hydro had been providing electrical power to downtown Winnipeg, at which time it was acquired by Manitoba Hydro.

Manitoba Hydro is a Crown Corporation regulated by the Province of Manitoba and a fully integrated electrical utility with large-scale generation, transmission, and distribution operations. Manitoba Hydro can therefore consider the total system cost and benefits of new capacity development projects (most notably its massive hydroelectric dams), rather than, for example, relying on third parties for electricity supply and transmission.

In 2014, the Province of Manitoba commissioned a review of the management of Manitoba Hydro EE policies and the formulation of recommendations for possible alternatives. This led to the issuance of a report submitted to the Public Utilities Board on the circumstances that led to the development of a demand-side approach to EE management.

"In the 1960s and 1970s, utility planning focused primarily on assessing supply options to meet demand forecasts. Over time, this approach was discredited because it took demand as a given. To the contrary, utilities and regulators realized that demand could be "shaped," through DSM efforts, in much the same way that supply could be built. As a result, planning evolved away from 'Which combination of supply resources is the cheapest way to meet forecast demand?', and toward 'Which combination of supply or demand resources is the least costly and least risky way to achieve equilibrium?', i.e. to keep the lights on. This came to be known as Least-Cost Integrated Resource Planning, or IRP for short."[2]

"Least-Cost Integrated Resource Planning (IRP) stems from as far back as the 1960s, when arguments were brought up in favor of competitive provision of power generation. The first energy crisis of the 1970s accelerated thinking about resource planning and risk, as utilities faced massive cost overruns and, just as critically, lower-than-projected demand, costing billions of dollars of ratepayers' money. The next decades saw the development of new supply-side technologies and options, as well as increasing efforts (and associated budgets) directed at DSM, adding further complexity to both the planning and energy procurement processes. IRP is a response to these multiple challenges. Its goal is to minimize the total societal cost of energy generation—and use—over the long term. It does so by seeking to evaluate all potential resources—both supply- and demand-side—on an equal footing and in a timely manner. Resources can include:

- On the demand side: EE programs, demand response initiatives, direct load control, interruptible power, rate structure changes, demand-side renewables (e.g. solar PV), industrial cogeneration, behavioral encouragement programs, fuel switching and fuel retention programs, conservation voltage regulation, T&D[i] efficiency improvements, and others as well; and

- On the supply side: A variety of technologies (intermittent renewables, baseload renewables, nuclear, or fossil-fired plants; and "peaker" plants) and strategies (e.g. utility-owned, PPAs with independent power producers, PPAs with other utilities, purchases from short-term energy and capacity markets, and increased transmission capacity) to increase imports and load balancing."[3]

April 1991: Manitoba Hydro Introduces the Power Smart Program
Three years after B.C. Hydro introduced its Power Smart program, Manitoba Hydro adopted its own version of the energy conservation program.

In a newsletter from 1991, Manitoba Hydro CEO explained that "Power Smart was selected as a means of conveying the concepts of DSM to Manitoba Hydro's customers. It will be used by the Corporation to reduce the growth of the province's electrical consumption in a manner that is cost-effective, is sustainable, serves our customers' needs and meets public concern for the environment."[4]
By introducing Power Smart to Manitobans, Manitoba Hydro sought to achieve four major goals:

1) Encourage manufacturers and retailers to produce and promote energy efficient products and improve the availability and pricing of energy efficient appliances and energy saving devices;

2) As part of a national Power Smart effort, enable nationwide advertizing and promotion by Canada's electric utilities on a scale larger

---

[i] Transmission and distribution (T&D)

than the promotional budgets of utilities acting alone;

3) Utilities would be able to act in conjunction with manufacturers and government agencies to develop standards and improve energy efficient applications by pooling existing knowledge and research funds; and

4) Utilities would be able to work together and share information on effective versions of the Power Smart program, thus improving customer acceptance of new technologies and techniques while minimizing program administration costs for utilities.[5]

Demand-side Management (DSM)

Manitoba Hydro's demand-side management strategy, the Power Smart DSM initiative marketed under the Power Smart brand, is designed to encourage the efficient use of energy in the residential, commercial, and industrial customer sectors.

Manitoba Hydro's overall DSM strategy involves adopting a broad approach to capturing EE opportunities: education to build awareness and understanding about EE, motivating customers with the aid of financial tools, and entrenching energy savings by supporting the adoption of federal and provincial codes and regulations.[6]

Manitoba Hydro also encourages initiatives to reduce energy consumption and energy bills by providing energy auditing, education, incentives and energy alternatives to residential, commercial and industrial customers. The program is designed to attract and retain customers who value the province's energy supply, move the province closer to GHG reduction targets, provide lower-income ratepayers with affordable upgrades such as high-efficiency furnaces and insulation and create new jobs in sustainable areas.

In assessing options for pursing DSM opportunities, Manitoba Hydro uses a number of metrics to assess EE potential. These metrics assist in determining whether to pursue an opportunity, how aggressive it will be pursued, the effectiveness of program design options, and the relative investment sharing between ratepayers and participating customers. These metrics include the Total Resource Cost, Societal Cost, Rate Impact Measure, Levelized Utility Cost, and Customer Simple Payback. In addition to conducting quantitative assessments, Manitoba Hydro also considers various qualitative factors including equity (i.e. reasonable participation by various ratepayer sectors such as lower income) and overall contribution of the EE opportunities toward a balanced energy conservation strategy and plan.[7]

Key Focus Areas of Manitoba Hydro
- Develop, implement and promote Power Smart programs that aggressively pursue all economic EE opportunities for both electric and natural gas customers;
- Capture EE improvements through the advancement of codes and standards, service extension policies and rate structures;
- Promote the optimum use of electricity, natural gas and alternative renewable energy sources (RES);
- Assess customer load displacement options.

STAKEHOLDERS

Manitoba Public Utilities Board
      The Manitoba Public Utilities Board (PUB) is an independent quasi-judicial administrative tribunal operating under the authority of the Manitoba Legislature. While the current Public Utilities Board Act was passed in 1959, the PUB has regulated services under other legislation since 1913. The PUB has broad oversight and supervisory powers over public utilities—including Manitoba Hydro and Centra Gas—and any other designated monopolies, as set out in statute. The PUB considers both the impact to customers and financial requirements of the utility when approving rates.

Centra Gas Manitoba
      Centra Gas Manitoba Inc. distributes, stores, and transports natural gas in Manitoba. The company was formerly known as ICG Utilities (Manitoba) Ltd., but changed its name to Centra Gas Manitoba Inc. in January 1991. In 1999, Manitoba Hydro acquired Centre Gas and, since then, most Manitobans have obtained both their electricity and gas from Manitoba Hydro.

Sustainable Building Manitoba
      Sustainable Building Manitoba Inc. was incorporated in 2005 as a nonprofit organization. Since that time, the organization has focused on its vision of "A sustainable built environment in Manitoba." Its activities in networking and education have made it the most recognized sus-

tainable building knowledge hub for governments, media and industry partners alike.

Consumers' Association of Canada—Manitoba Branch

The Manitoba branch of the Consumers' Association of Canada (CAC Manitoba) is an independent, nonprofit, volunteer organization working to inform and empower consumers. It also represents consumer interests in Manitoba.

Established in 1947, CAC Manitoba is governed by an elected board of directors. Through representation on boards and committees, presentations to committees and at rate hearings, as well as by other means, CAC Manitoba puts forward the consumer point of view on safety, prices and quality of service in the energy sector.

Green Action Centre

The Green Action Centre is a nonprofit and non-governmental organization serving Manitoba. Its mandate is to promote greener living through environmental education and by encouraging green solutions for households, workplaces, schools and communities. It promotes practical measures to improve the sustainability and quality of life of Manitobans. Research, education and advocacy activities are addressed to individuals, institutions, businesses and governments, including the PUB and Manitoba Hydro.

POLICIES AND STRATEGIES

Manitoba Energy Efficiency Policy

The Manitoba EE policy is overseen by the Energy Division of Manitoba Growth Enterprise and Trade. EE is the quickest, cleanest and least costly way for Manitoba to meet its growing energy demands. EE also contributes to several other important economic, social and environmental goals in Manitoba, such as:

- Reducing energy costs for individual consumers;

- Expanding ability to export clean, renewable electricity from local hydroelectric and wind sources;

- Increasing energy security by reducing need to import fossil fuels;

- Allowing deferral of additional new generation;
- Improving economic competitiveness;
- Addressing energy affordability concerns for low and modest-income households;
- Creating new opportunities for local businesses and employment;
- Reducing GHG emissions.

Manitoba Climate Change and Green Economy Action Plan (2015)

Released in December 2015, the Manitoba Climate Change and Green Economy Action Plan committed Manitoba to adopting a cap-and-trade program for large industrial emitters to be linked with the Ontario and Quebec cap-and-trade system. However, Manitoba's current Premier, Brian Pallister, stated in late 2016 that the province would not be proceeding with a cap-and-trade program, but would consider implementing a carbon tax. Other commitments in the plan, including GHG emission reduction targets, are equally uncertain given the province's evolving approach to climate change mitigation.

Carbon and GHG Legislation in Manitoba

As of 30 June 2014, Manitoba imposed an emissions tax on petroleum coke (petcoke) used for industrial purposes through the Emissions Tax on Coal and Petroleum Coke Act. Revenues from the tax were to contribute to programs that help facilitate the conversion to renewable energy sources, primarily biomass. The province also began to phase in a ban on petcoke and coal for heating purposes on 1 January 2014, and full compliance was required by 1 July 2017.

LEGAL AND REGULATORY FRAMEWORK

The Energy Act—1994

This act established the Energy Department as the agency responsible for energy planning and policy development, particularly on energy supply and demand issues. The Energy Department is responsible for the development and coordination of contingency plans to deal with possible energy shortages. The act also provides for the appointment of an advisory committee to respond to any energy matters which may require public consultation or advice.

The Climate Change and Emissions Reductions Act—2008

The purpose of this act was to address climate change, encourage and assist Manitobans in reducing emissions, set targets for reducing emissions and promote sustainable economic development and energy security. The initial reduction target for Manitoba was to reduce emissions by 31 December 2012 to at least 6% below total 1990 levels.

The Energy Savings Act—2013

When the Province of Manitoba passed Bill 24, the Energy Savings Act, it was the first legislation of its kind in North America. It made EE improvements accessible to all Manitobans through an innovative on-meter financing mechanism and created the potential for poverty reduction, community renewal and social enterprise development.

Efficiency Manitoba Act—2017

The purpose of this Act is to establish Efficiency Manitoba as a corporation with a mandate laid out in Section 4: establish savings targets for Efficiency Manitoba relative to electrical energy and natural gas consumption, as well as establish a funding and regulatory oversight framework for Efficiency Manitoba.

The mandate of Efficiency Manitoba is to: (a) implement and support demand-side management initiatives to meet savings targets and achieve any resulting reductions in GHG emissions; (b) achieve additional reductions in the consumption of electrical energy or natural gas, including reductions in electrical power demand if they can be cost-effectively achieved; (c) mitigate the impact of rate increases and delay capital investments in major new generation and transmission projects; (d) if any of the following are prescribed as being subject to demand-side management under this Act, carry out the prescribed duties in respect of these—demand for electrical power, consumed potable water, as well as consumed fossil fuels in the transportation sector; and (e) promote and encourage the involvement of the private sector and other non-governmental entities in the delivery of provincial demand-side management initiatives.

The Manitoba Energy Code for Buildings (MECB)

The purpose of the Manitoba Energy Code for Buildings (MECB) is to reduce building energy use. The MECB is estimated to render buildings 25% more energy efficient. The MECB became effective 1 December 2014 and supplements the Manitoba Building Code 2011 (MBC). It applies to any new buildings and new additions that are referred to as Part 3 Buildings under the MBC, which are generally buildings larger than 600 square meters in floor area.

EE Standards—Minimum Energy Performance Standards (MEPS)

The Energy Act (1994) includes Energy Efficiency Standards for Replacement Forced Air Gas Furnaces and Small Boilers Regulation which established new minimum energy performance standards for replacement gas furnaces and small boilers. The efficiency requirement set for gas furnaces is 92%, the highest level in any North American jurisdiction.

Carbon Tax

After the federal government released details of its plan for carbon pricing in May 2017, Manitoba's Conservative government indicated that more consultation would be required before the province could share its plans with the public. An initial price of $10 per ton was set by Ottawa on carbon dioxide emissions for 2018, which increases to $50 per ton in 2022. This represents an extra 2.33 cents per liter at the pump, and 11.63 cents per liter by 2022. The Manitoba Premier has not ruled out eventually increasing costs for carbon-emitting fuels, which opens the possibility of redirecting the revenue derived from a carbon tax to Manitoba Hydro, thereby improving the financial situation of the provincial Crown Corporation and helping to keep electricity rate increases to a minimum.

FINANCIAL MECHANISMS

Incentive Programs and Financing

The Government of Manitoba provides dozens of EE incentives and financing programs. For example, Manitoba Hydro offers loans and financing for the purchase of new appliances, while also offering a rebate to homeowners who retire certain old and inefficient appliances. Under this program, representatives pick up old inefficient refrigerators and hand each participant a $50 check.

Manitoba Geothermal Energy Incentive Program

The province encourages homeowners to adopt the use of geothermal energy with a refundable tax credit that is retroactive to April 2007 and a provincial grant program that became effective in early 2009. For geothermal installations in new homes, incentives of up to $3,000 are available to Manitoba residents. For existing home conversions to geothermal, Manitobans are eligible to apply for a Green Energy Tax Credit of up to $2,000.

Power Smart Solar Energy Program

In 2016, Manitoba Hydro launched the Power Smart Solar Energy Program, thereby enabling home and business owners to generate their own power with solar panels and sell excess electricity to Manitoba Hydro.

A full list of Manitoba Hydro Power Smart incentive and financing programs, from the Affordable Energy Program (for lower income households) to the Water & Energy Saver Program, are available on the Manitoba Hydro website.

Chapter 5

New Brunswick

ENERGY SECTOR

New Brunswick energy consumption decreased by 18% from 1995 to 2015, mostly due to a 21% decline during the 2010-2015 period. Most of this decline is attributed to the industrial sector (493 PJ) and the transportation sector (104 PJ). The energy consumption of all other sectors also decreased, except for residential which increased by 23% from 1995 to 2015.

In 2013-15, the transportation sector consumed the most (38%), followed by the industrial (26%) and the residential (20%) sectors. The commercial sector consumed 10% of total energy consumption, while the agriculture and public administration sectors combined for 5% of total consumption for the three-year period.

Figure 5-1 illustrates average energy consumption in three-year periods from 1995 to 2015.[1]

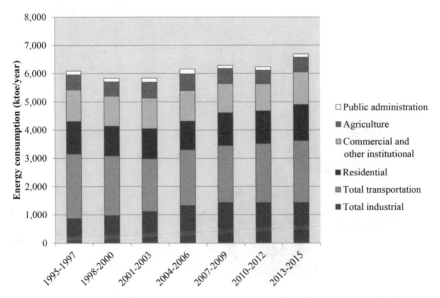

Figure 5-1: New Brunswick Energy Consumption from 1995 to 2015

ENERGY EFFICIENCY BACKGROUND

The three major sources of energy consumed in New Brunswick (NB) are petroleum products (123 PJ), electricity (46 PJ) and natural gas (22 PJ).[2] Table 5-1 summarizes fuel use by sector.

Table 5-1: Energy Consumption in NB by Fuel Type and Major Sectors (2015)

	Industrial	Transport	Agriculture	Buildings
NG	79.9%	0%	0%	20.1%
Electricity	29.1%	0%	0.5%	70.5%
Petroleum	8.7%	69.8%	3.8%	17.7%

In 2015, three quarters of consumed electricity was produced within the province. The province's installed capacity mix was 4,251 MW as illustrated in Figure 5-2.[3]

NB Power Corporation (NB Power), a Crown Corporation owned by the government of New Brunswick, is the primary electrical utility in New Brunswick. It produces, transmits and distributes electricity to all energy consumers except those in three municipalities served by local

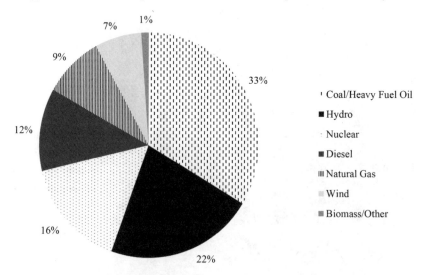

Figure 5-2: Installed Capacity Mix for Electricity Production (2014)

electrical utilities. Most days, a combination of NB generation sources along with imports from New England and Quebec supply power to NB Power clients.

Enbridge Gas NB distributes natural gas and is the sole franchisee for natural gas distribution in the province.

A Brief History

Since the 1980s, as a result of the oil crisis of the 1970s, NB has implemented EE measures to reduce energy demand. In the 1990s, NB energy policies integrated EE goals based on a study that had determined the EE potential of the province.[4] The study analyzed economically attractive EE opportunities in the residential, commercial, industrial and transportation sectors. Results indicated that gross energy savings could attain 37% in residential, 22% in commercial, 26% in industrial and 13% in transportation. Achievable energy savings, if market barriers were overcome, were estimated at approximately one third of said gross potential. Given the NB electricity production mix, most electricity is inefficiently produced from fossil fuels. The study also stated that fuel switching from electricity to oil had great potential for improving primary energy use and overall economic efficiency. The study also evaluated non-energy benefits. The study claimed that more than 2,000 jobs would be created annually due to the increase in disposable income and multiplier effects if the province undertook fuel-switching. It was also estimated that EE measures could lead to significant reductions in CO_2 (7.7 million tons), NO_x (35,200 tons) and SO_2 (98,000 tons) emissions.

In 2001, the Government of NB released the White Paper, New Brunswick Energy Policy 2000-2010, a 10-year energy policy articulated around five goals aimed at ensuring a secure, reliable and cost-effective energy supply for energy users across the province.[4] The energy policy also offered direction on issues such as EE and alternative energy sources. EE was considered a key component in attaining NB energy policy goals. Given the significant economic and environmental benefits of EE, the energy policy identified the need to develop and implement a comprehensive EE strategy.

In 2005, the Energy Efficiency and Conservation Agency of New Brunswick (Efficiency NB) was established as a Crown Corporation to act as the primary organization for the promotion of EE and conservation in NB. Efficiency NB promoted EE in the residential, commercial, public and industrial sectors by developing and delivering programs and initiatives in line with its mission. Throughout its existence, Efficiency NB delivered programs for existing and new residential homes and commercial buildings, as well as industrial facilities. Due to budget cuts, Efficiency NB suspended all new construction and industrial programs. In its last year of operation, Efficiency NB had an approximate budget of $12 million.[5]

In 2014, Efficiency NB was dissolved and activities and employees of the former agency were integrated into NB Power.[6] As a result, the Efficiency Services Division of NB Power was created and the integration of Efficiency NB was finalized in fiscal year 2015-2016. The Efficiency Services Division focuses on the promotion of EE and climate change initiatives.[7]

The Government of NB established an Energy Commission in 2010 to consult the population on how to change the energy regulation for the benefit of the economy and society. As a result of these consultations and other work, the NB Government released, in 2011, the New Brunswick Energy Blueprint which provided a 10-year vision for the energy sector. The Blueprint also outlined a three-year action plan comprised of 20 action items. Three of these action items directly related to EE: Energy Efficiency Building Code Standards; Energy Efficient Appliances; and Equipment and Electricity Efficiency Plan.[8] In August 2014, the Final Progress Report of the New Brunswick Energy Blueprint proclaimed that all 20 actions items had been achieved, including the Electricity Efficiency Plan.

The Electricity Efficiency Plan consisted of requiring electric utilities, in conjunction with Efficiency NB (which still existed in 2011), to release a three-year electricity efficiency plan. In 2012, a steering committee was established to outline the initial Electricity Efficiency Plan covering the 2013-14 to 2015-16 period. The committee included senior officials from Efficiency NB, the Department of Energy and Mines, the Department of Environment and Local Government, as well as the electrical utilities (NB Power, Saint John Energy, Edmundston Energy and Perth-Andover Electric Light Commission). The plan allocated a $57 million investment in EE programs for the 2014-2017 period.[9] Given that Efficiency NB was integrated into NB Power in 2014, NB Power completed and released the three-year Electricity Efficiency Plan in cooperation with the three NB municipal electric utilities.

In response to the Paris Agreement on Climate Change (COP21) drafted in 2015, whereby Canada agreed to contribute to maintaining increases in global average temperatures below 2°C, the First Ministers launched a nationwide process consisting of developing a pan-Canadian framework on clean growth and climate change. Seen as an opportunity for NB "to be more efficient and competitive, to open new business opportunities and to build more resilience into our aging infrastructures," the Government of NB released the 2016 Climate Change Action Plan: Transitioning to a Low-Carbon Economy. This plan outlines the actions of the province to reduce GHG emissions and recognizes that "energy we do not use is free of cost and emissions." Seven of the 118 NB climate actions included in the Climate Change Action Plan relate to EE, including: mandate EE delivery agents to provide EE initiatives; increase spending on EE in the capital budget by 50%; use the Property-Assessed Clean Energy (PACE) program, if viable; adopt the latest National Energy Code of Canada for Buildings and National Building Code; and require energy labeling for all new building constructions.

Figure 5-3 presents the initiatives that fostered EE development in NB.

STAKEHOLDERS

The Department of Energy and Mines is in charge of energy legislation in NB. The department pursues the following EE objectives, as per the New Brunswick Energy Blueprint, 2011: "Low and stable energy prices; Energy security; Reliability of the electrical system; Environmental responsibility; Effective regulation."[8]

The New Brunswick Department of Energy and Mines "administers the Electricity Act and Reliability Standards Regulation which establishes the authority and requirements for the adoption and enforcement of electric reliability standards" in the province.[10]

The New Brunswick Energy and Utilities Board (EUB) is an independent Crown Agency. It regulates electricity and natural gas utilities, as well as motor carriers. The EUB therefore works with representative groups to ensure the public interest.[11]

NB Power is a Crown Corporation wholly-owned by the Government of NB and is responsible for electricity generation, transmission and distribution. As of 2017, NB Power had 12 generating stations totaling a net electrical production capacity of 2,853 MW servicing close to 400,000 customers. Three other municipally-owned electrical utilities

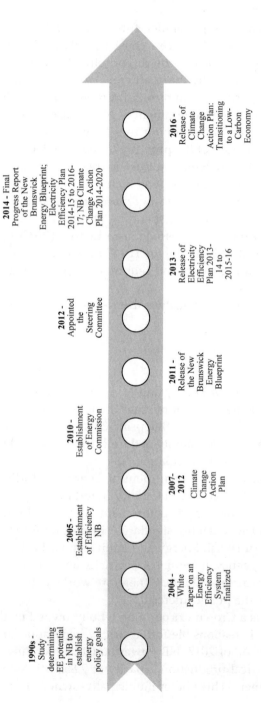

Figure 5-3: Timeline of EE Initiatives in NB

operate in the province: Edmundston Energy; Perth Andover Electric Light Commission; and Saint John Energy. Since 2014, NB Power has been responsible for promoting EE and climate change initiatives in the electricity sector.

A subsidiary of NB Power, the New Brunswick Energy Marketing Corporation is in charge of electricity trading outside the province.[14]

POLICIES AND STRATEGIES

The EE strategy for the province, White Paper on NB Energy Policy, was released in 2000 and established an energy policy framework up to 2010. The energy policy had the following goals:[12]

• Ensure a secure, reliable and cost-effective energy supply for residential, commercial and industrial customers;

• Promote economic efficiency in energy systems and services;

• Promote economic development opportunities;

• Protect and enhance the environment;

• Ensure an effective and transparent regulatory regime.

Another important document for EE and energy conservation in NB is the New Brunswick Energy Blueprint released in 2011,[8] a long-term vision and a strategic action plan for the sector.

Table 5-2 lists all actions included in the Blueprint along with the completion year.

In response to the actions in the NB Energy Blueprint, a certification program was launched called Energy Fundamentals for Leaders at the University of New Brunswick Saint John Campus. The program targets professionals in the energy sector who would like to gain in-depth knowledge of energy issues.[14]

An outcome of the actions of the NB Energy Blueprint was the release of the Climate Change Action Plan in 2014. After the Paris Agreement on Climate Change, an updated version was released entitled Transitioning to a Low-Carbon Economy—New Brunswick's Climate Change Action Plan. This update outlines the vision for reducing GHG emissions while striving for economic growth.

The plan contains actions in the following areas:

Table 5-2: Actions Established in the Blueprint and Year of Completion[i], [8], [13]

Action	Energy Action Plan	Completion Year
1	Reintegration of NB Power – merger of NB Power companies into a single vertically integrated Crown utility.	2013
2	Electricity Market and NB System Operator - review of electricity market policies and implementation of appropriate structural and operational changes, including the dissolution of the NB System Operator and migration of system operator functions back to NB Power.	2013
3	NB Power – Debt Management Plan – implementation of a debt management plan for NB Power, allowing the utility to reduce debt and create shareholder equity.	2012
4	NB Power – Regulatory Oversight and Integrated Resource Plan - all NB Power operations are subject to regulatory oversight and review. Also, NB Power presents an Integrated Resource Plan every three years and an annual Financial Forecast to EUB, or as directed by the EUB.	2013
5	Regional Electricity Partnerships – pursuing regional electricity agreements, joint ventures and partnerships where there are positive commercial outcomes for NB Power and defined benefits for NB ratepayers.	2012
6	Smart Grid Technology and Innovation – expansion of the network of smart grid stakeholders and partnerships and work with existing and new smart grid pilot projects.	2012
7	Large Industrial Renewable Energy Purchase Program – to bring the electricity costs of qualifying large industrial companies in line with their Canadian competitors by implementing a Large Industrial Renewable Energy Purchase Program.	2012
8	Renewable Portfolio Standard - increase the provincial Renewable Portfolio Standard to a minimum of 40% of NB Power's in-province sales by 2020.	2014
9	Future Development of NB's Renewable Energy Resources: (a) support local and First Nations small-scale renewable projects; (b) integrate current and future wind generation in the most cost effective and efficient manner; (c) support promising solar, bio-energy and other emerging renewable energy technologies.	2014
10	Wood Based Biomass Resources - development and implementation of support policies to optimize the energy output of NB wood based biomass resources with a specific focus on pellets.	2014
11	Energy and Climate Change – development of the key energy components for the 2012-2020 NB Climate Change Action Plan.	2014
12	Electricity Efficiency Plan - electric utilities, in conjunction with Efficiency NB, preparation of a three-year electricity efficiency plan.	2014
13	Energy Efficiency Building Code Standards – requirement for minimum EE standards for new building constructions in NB by adopting national standards and amend the New Brunswick Building Code Act to create the authority to do so.	2014
14	Energy Efficient Appliances and Equipment – upgrading the list of regulated appliances and equipment under the Energy Efficiency Act.	2012
15	Natural Gas Distribution Rates – review of the natural gas rate structure and distribution network to achieve a cost-based rate structure and improve access to natural gas across the province.	2012
16	Petroleum Products Pricing - review of the Petroleum Products Pricing Act and Regulations to ensure continued effectiveness in achieving the objectives of petroleum price stability while ensuring the lowest possible price to the consumer without endangering continuity of supply.	2014
17	Energy and Utilities Board - restructure the Energy and Utilities Board to consist of exclusively full-time members.	2013
18	Office of the Public Energy Advocate - establish and appoint a full-time public energy advocate to replace the system of ad hoc appointments of public intervenors.	2013
19	Energy Literacy, Education and Skills Development: (a) develop an energy sector workforce development strategy; (b) pilot an energy certificate program; (c) implement an energy literacy, education and awareness program.	2014
20	Energy Research and Development - development and implementation of a NB energy sector research and development strategy supporting the adoption of emerging clean energy technologies.	2013

[i] Table 5-2 is drawn from "The New Brunswick Energy Blueprint Final Progress Report, 2014" (page 1); clarification of each action is cited from "The New Brunswick Energy Blueprint, 2011."

- "Provincial government leadership;

- Collaboration with First Nations;

- GHG emission reductions;

- Adaptation to the impacts of climate change;

- Economic opportunities;

- Accountability and reporting;

- Funding for climate change."[15]

LEGAL AND REGULATORY FRAMEWORK

The General Regulation—Energy Efficiency Act specifies obligations on EE and energy conservation in NB.[16] In 2012, a list of regulated appliances and equipment with minimum efficiency levels was included in this law under a commitment made in the NB Energy Blueprint.[13] This list aims to allow consumers to make better and more informed product choices. It is updated every two years because the efficiency levels of new products continue to rise. The EE Act is an important complement to the EE standards in the NB Building Code. Another action, #13 Energy Efficiency Building Code Standards, established in the Blueprint led to the NB government adapting the National Building Code of Canada 2010 which was then integrated into the NB Building Code Act that came into force in 2015.

In 2016, the federal government announced that each province needed to implement a carbon price mechanism and pay a price on carbon emissions of $10 per ton as of 2018. Therefore, based on the NB 2014 Climate Change Action Plan, the NB implementation mechanism adheres to the following stated principles: "The provincial government will take into consideration impacts on low-income families, trade-exposed and energy-intensive industries, and consumers and businesses, when developing the specific mechanisms and implementation details, including how to reinvest proceeds. Any carbon pricing policy will strive to maintain competitiveness and minimize carbon leakage (i.e. investments moving to other jurisdictions). Proceeds from carbon emission pricing will be directed to a dedicated climate change fund."[15]

In December 2015, NB Power released its first DSM plan for the
2016-2018 period. The plan introduced DSM initiatives in the residential, commercial and industrial sectors.

FINANCIAL MECHANISMS

According to the 2014/15-2016/17 Electricity Efficiency Plan,
investments in EE programs will total nearly $57 million, resulting in
estimated 48,100 tons of CO_2 emission reductions and $80 million in bill
savings over the life of the implemented EE measures.[17]
Currently, the NB Power website offers five programs as follows:
• Home Insulation Energy Savings Program—This is a financial
  incentive and retrofit program for the residential sector to compensate homeowners for insulation and air sealing upgrades installed
  in their home. An initial home evaluation is carried out to determine the measures to be implemented.
• Low-income Energy Savings Program—This program is available
  for low-income homeowners who need major efficiency upgrades.
  It aims to help reduce electricity bills. Eligible measures include:
  compact fluorescent lighting, domestic hot water pipe insulation;
  low-flow showerheads, insulation (basement/crawlspace, attic,
  main wall), faucet aerators, ductless heat pumps, and air sealing.
• Commercial Buildings Retrofit Program—This program is available to the commercial and public sectors and provides financial
  incentives ranging from $3,000 to $75,000.
• Small Business Lighting Program—This program provides financial incentives to compensate small businesses for the costs of
  lighting replacements or lighting system and control upgrades.
• Industrial Energy Efficiency Program—This program aims to facilitate the implementation of EE measures in the industrial sector. It
  provides financial incentives or advice.[18], [19]

The 2014/15-2016/17 Electricity Efficiency Plan requires that, as
a future step, the NB Power plan integrate a smart-grid system with
the aim of facilitating the implementation of future programs to reduce
demand. The focus will be on maximizing the effect of the initiatives
described in the plan.

Chapter 6

Newfoundland and Labrador

Ms. Krista LANGTHORNE, Newfoundland Power

ENERGY SECTOR

Over the last 20 years, annual energy consumption in Newfoundland has fluctuated slightly. For the 2013-2015 period, average annual consumption was 2,702 ktoe, 2% more than the 1995-1997 average. Although total consumption remained almost constant between 1995 and 2015, the consumption of every sector varied significantly. Commercial sector energy consumption increased by 70% from 1995 to 2015, and this sector consumed 12% of total consumption in 2013-15. The other large increase occurred in the transportation sector, a 30% increase to reach 42% of total consumption in 2013-15. Also over this three-year period, the residential sector accounted for 18% of provincial energy consumption.

The industrial sector was the second most energy intensive with 25% of total consumption, diminishing consumption by 24% over the last 20 years. Public administration consumption fell by more than half with a 62% decrease in 2013-15 compared to 1995-97. The latter represented 3% of provincial energy consumption, while agriculture accounted for only 0.3%.

Figure 6-1 illustrates average annual energy consumption in periods of three years from 1995 to 2015.

ENERGY EFFICIENCY BACKGROUND

History of Conservation in Newfoundland

EE became public policy in Newfoundland and Labrador in the late 1980s and early 1990s. Newfoundland Power and Newfoundland and Labrador Hydro (the Utilities) became the two major EE delivery agents in the industry. EE became a central issue during a hearing for Newfoundland Power's rate application in 1989. Environmental concern was growing as a result of thermal generation plants and the use

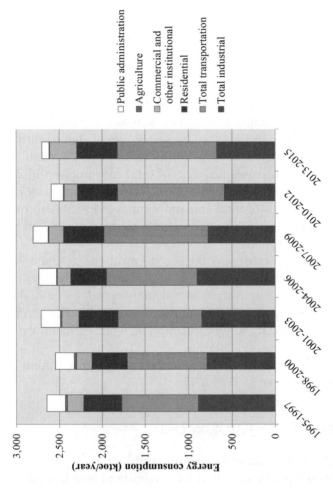

Figure 6-1: Newfoundland and Labrador Annual Consumption from 1995 to 2015
Source: Statistics Canada, CANSIM 128-0016

of land for power system facilities. The provincial regulator, the New-foundland and Labrador Board of Commissioners of Public Utilities, also recognized that conservation and EE were important issues that needed to be factored into the Utilities' least-cost planning.[i] Available demand-side alternatives had to be integrated with supply-side alter-natives to determine which combination of such options would provide the most economic and environmentally sensitive plan to meet custom-er demand and eco-concern expectations.

Both utilities began offering a range of residential and commercial conservation programs in 1991. Program offerings covered low-flow showerheads, commercial lighting, water heaters, attic and basement insulation, and thermostat upgrades.

While these new conservation programs were being developed and implemented, the Canadian economy fell into recession and the Newfoundland economy was experiencing the first impacts of the fish-ery moratorium. Changes in the economy created the possibility of low or negative load growth which would necessitate energy rate increases, thereby further decreasing load growth. Conservation and EE during this period of low load growth contributed to less efficient utilization of exist-ing facilities and many customer conservation programs were deferred.

In 1992, the Provincial Economic Recovery Commission created a not-for-profit agency to promote conservation, the Conservation Corps of Newfoundland and Labrador (CCNL). The purpose of CCNL was to increase quality employment during the economic downturn. CCNL promoted environmental and cultural enhancement through training and career development opportunities for youth and community part-ners throughout the province.

While two programs for attic and basement insulation and ther-mostat rebates were still being offered by the Utilities, conservation activities focused mainly on customer education. This continued until the mid-2000s when load growth led to forecasted energy and capacity constraints on the island. The provincial regulator intervened by intro-ducing a new demand-energy rate.[ii] This new rate was intended to al-low the Utilities to provide incentives through energy conservation and demand management (CDM) programs intended to lower long-term

---

[i] Order No. P.U. 1 (1990) issued January 30, 1990, by the Board of Commissioners of Public Utilities.

[ii] P.U. 44 (2004) issued December 8, 2004, by the Board of Commissioners of Public Utilities.

consumer electricity costs.

A conservation potential study was conducted by the Utilities in 2007, the results of which were used as direct input in the Utilities first joint five-year plan. The plan outlined potential technologies, programs, supporting elements, and cost estimates that supported the long-term goals of establishing a conservation culture and fostering sustainable reductions in electricity consumption. The Utilities followed up this plan by creating two more joint plans, the most recent being the Five-Year Conservation Plan: 2016-2020 that was filed with the regulator in 2015 and is currently being implemented. This plan builds on the experience of the Utilities and reflects the principles underlying the two previous plans.[iii]

The Government of Newfoundland and Labrador has worked cooperatively with the Utilities since the 1990s to ensure that conservation information and programs are available to all residents. This partnership has been primarily between the Utilities, the Department of Natural Resources, the Department of Municipal Affairs and Environment, and the Office of Climate Change (OCC) established in 2009.

As an example, The Natural Resources Canada Office of Energy Efficiency partnered with Newfoundland Power in 2004 to offer the Switch to Save initiative to help customers adjust to increasing electricity rates. The initiative encouraged customers to upgrade lighting to more efficient compact fluorescent bulbs.

Challenges of Conservation in Newfoundland and Labrador

In Newfoundland and Labrador, a number of factors make the delivery of EE programs unique. The Newfoundland electricity grid is an isolated electricity system commonly referred to as the Island Interconnected System (IIS). The IIS is not yet connected to the North American electricity grid. Consequently, its generation resources must be capable of meeting customer demand at all times throughout the year.

Two electrical utilities operate in the province. Newfoundland Power is primarily an electricity distribution and customer service organization, while Newfoundland & Labrador Hydro (NFL Hydro) is mainly an electricity generation and transmission organization. To ensure the cost effectiveness of the least-cost planning process, the Utilities

---

[iii] The Five-Year Energy Conservation Plan: 2008-2013 was filed with the Board on June 27, 2008. The Five-Year Energy Conservation Plan: 2012-2016 was filed on September 14, 2012.

offer energy conservation programs to customers on a joint and coordinated basis.

Approximately 60% of residents use electricity as their primary source of heating and the province's northern maritime climate means a long heating season. Therefore, the greatest EE potential is in reducing space heating. This has other consequences such as reducing the savings realized from other technologies as a result of interactive effects with the electric heat source. For example, the savings generated from installing an LED bulb must be discounted, typically by 60%, to account for the loss in the heat generated from a less efficient bulb.

The climate in Newfoundland and Labrador does not usually require air conditioning, especially in residential applications. Therefore, cooling savings claimed in other jurisdictions do not apply to the province. An example of this is heat pumps. The efficiencies generated from using a heat pump in the summer to provide cooling cannot be considered when assessing savings impacts. Rather, any cooling load would have a negative effect on savings because it adds to homeowner consumption.

The challenging geography of Newfoundland and Labrador renders offering certain types of customer energy conservation programs cost prohibitive. Programs such as direct install and home energy audit programs that rely on home visits quickly become too expensive because of the geographically isolated regions of the province, such that costs outweigh the benefits of energy savings.

Another unique characteristic of customer conservation programs in this province is the 21 isolated diesel systems on the coast of Labrador that are powered by diesel generators. Customers in these areas are charged different rates and have different cost savings economics that need to be considered distinct from the rest of the province. Nearly 90% of residents in this region heat using energy from fuels other than electricity.

The Utilities use the marginal costs of energy to evaluate the cost effectiveness of customer energy conservation programs. Significant changes to the IIS are anticipated in the near future. The Muskrat Falls hydroelectric facility is forecasted to be operational in 2019 and will connect the island for the first time to the North American grid. It is expected to reach nominal power production capacity by 2020. As a result, there is uncertainty as to the capacity and marginal costs of energy on the IIS, although these costs are expected to decline as the system is converted from thermal to hydroelectric generation. This uncertainty will negatively impact the cost effectiveness of customer energy conservation programs.

Interconnection costs are expected to be included in customer rates as of 2019, thus increasing electricity bills. This is expected to further motivate customers to manage their electricity costs and conserve energy. Also, the recent provincial economic slowdown is expected to positively influence customer behavior by fostering conservation efforts.

STAKEHOLDERS

The Utilities

Conservation and EE programs are mostly delivered by the electric Utilities in the province. The Utilities have offered customer energy conservation programs since 1991, as well as on a joint and coordinated basis under the takeCHARGE brand since 2009.

Newfoundland Power's primary business is electricity transmission and distribution throughout the island of Newfoundland and Labrador. This utility serves over 262,000 customers, representing 90% of all electricity consumers in the province.

Newfoundland Power purchases approximately 93% of the electricity it sells from NFL Hydro. Newfoundland Power generates the balance from its facilities, primarily smaller hydroelectric stations located across the island.

NFL Hydro, a Nalcor Energy company, is a Crown Corporation that is primarily an electricity generation company. NFL Hydro serves over 38,000 direct customers in rural Newfoundland and Labrador.

The Utilities have collaborated on customer energy conservation program planning and delivery for the past 25 years. Over the last nine years, programs have been offered jointly under the takeCHARGE brand. These programs include a variety of information dissemination and financial support activities to customers to assist them in managing energy usage. The energy conservation programs are responsive to customer expectations. They support efforts as responsible stewards of electrical energy resources and are consistent with the principle of providing least-cost and reliable electricity services.

To date, the primary focus of the Utilities' conservation initiatives has been the generation of energy savings. The secondary focus has been the development of a culture of conservation. These initiatives aim to address energy savings opportunities for customers in the residential, commercial and industrial sectors.

Government

The provincial government, primarily through the Department of Natural Resources, the Department of Municipal Affairs and Environment, and the OCC, is another important stakeholder in the EE marketplace.

The Department of Natural Resources is responsible for the stewardship and development of the province's natural resources through the Mines and Energy Branches.

The main objective of the Department of Municipal Affairs and Environment is to ensure that residents live in safe and sustainable communities. The department fosters environmental protection and enhancement by implementing water resource and pollution prevention regulations and policies, as well as coordinating environmental impact assessments.

The OCC is a central agency that leads policy and strategy development on climate change adaptation and mitigation issues, as well as EE. The OCC works collaboratively with other departments and agencies to ensure climate change and EE are effectively integrated into policy development and decision making.

The OCC also works actively with the federal government, other provinces and territories, industry, and stakeholders on policy matters related to climate change and EE. The OCC seeks new opportunities to move the province forward in these areas. It also assesses the implications of regional, national or international developments for the province.

The mandate of the OCC is to:

- Advance sustained action on climate change and EE that effectively balances economic and environmental considerations, including deepening public awareness, understanding, and engagement.

- Undertake focused research and analysis to enable the province to maximize opportunities, minimize risks from the impacts of climate change, and transition toward a lower-carbon global economy.

- Work with departments to better integrate climate change and EE considerations into current and future programs, services, legislation and regulations, and ensure effective coordination across government.

•   Advance the province's interests and priorities in regional, national and international forums on climate change and EE, as well as engage external stakeholders to deepen and widen government dialogue on next steps.

Non-Profit
The CCNL is a not-for-profit organization that focuses on conservation and the enhancement of environmental and cultural assets of the province through employment and outreach. It operates youth employment programs with funding from various sources such as the Department of Advanced Education and Skills, Canada Summer Jobs, corporate sponsors, municipalities, and community organizations.

The Newfoundland and Labrador Environmental Industry Association (NEIA) is a not-for-profit association created in 1992 that promotes the development of the green economy in the province. It is a resource for the environmental sector and offers a range of knowledge and support services for businesses working to increase their economic opportunities while respecting the environment.

POLICIES AND STRATEGIES

Utility Strategies and Plans
The Utilities developed multi-year conservation plans through the takeCHARGE partnership in 2008 and 2012. These plans included an overview of the current provincial conservation marketplace and outlined the strategy to be adopted for joint conservation activities. The plans also provided opportunities for customers to cost-effectively manage energy usage. Thus, joint customer energy conservation education and outreach activities were developed by establishing new programs for both the residential and commercial sectors.

The Utilities are currently implementing their *Five-Year Conservation Plan: 2016-2020*. This most recent plan builds on experience gained thus far and continues to reflect the principles underlying two previous joint plans. It further refines the opportunities identified in a recently updated conservation potential study that includes in-depth local market research and program cost benefit analysis.

Trade ally and customer outreach activities are a significant component of the Utilities' efforts to educate customers about energy conservation. The takeCHARGE team regularly advises the public and special interest groups by participating in presentations and tradeshows across the island. Resources include an information booth and interactive displays to attract visitors and foster interest.

Table 6-1 presents both actual and forecasted energy savings based on values drawn from the 2016-2020 plan.

Table 6-1: Conservation Program Five-Year Conservation Plan Energy Savings Estimates: 2009-2020 (F) in (GWh)

	2009-2012	2013-2015	2016-2020 (Forecast)	Total
Energy Savings (GWh)	66.9	232.2	894.7	1,193.8

Government Policy

The 2007 Energy Plan: Focusing Our Energy, developed by the Department of Natural Resources, describes the provincial vision for energy resource development. It established the government strategy to increase the use of energy resources in an economically and environmentally sustainable manner. It also expounded the importance of EE and conservation to the environment and the provincial economy.

A key commitment in this Energy Plan was the development of EE and conservation programs that not only generated savings for residents, but also protected and preserved the environment. As a result, the government launched two programs in 2009.

The Newfoundland and Labrador EnerGuide for Houses Program provided $300 toward the cost of a complete home EE audit and increased the federal ecoEnergy Retrofit Program grant by an additional $1,500 toward home efficiency improvements. The goal was to make EE home improvements more affordable and help homeowners reduce their energy costs. The program ran until 2011.

The Residential Energy Efficiency Program (REEP) was available to residents with a household income of $32,500 or less who use either fuel or electricity as a heating source. The program was administered through Newfoundland Labrador Housing. REEP covered the cost for pre and post-energy inspections and offered efficiency improvement grants up to a maximum of $3,000 per house in Newfoundland and $4,000 per house in Labrador. The program ran until 2017.

Another initiative includes the Newfoundland and Labrador Green Fund established by the government in 2007, the purpose of which is supporting projects that provide net reductions in GHG emissions. Funding of up to $25 million is offered by both the provincial and federal governments.

Furthermore, the Government of Newfoundland and Labrador developed the Build Better Buildings (BBB) Policy which came into effect in September 2010. The purpose of this policy is to establish the parameters under which future provincially funded infrastructure projects are to be carried out. The policy requires that new infrastructure projects and renovations over a certain cost and larger than a certain size be completed sustainably. These projects need to exceed the 1997 MNECB by 25%.[iv] Fulfilling this requirement usually entails improved insulation, efficient windows and lighting, as well as heating, ventilation and air conditioning (HVAC) system upgrades.

Another BBB requirement is that new builds must achieve a minimum standard under the Leadership in Energy and Environmental Design Program (LEED). Projects must register with LEED Canada and achieve a minimum of LEED Silver Certification.[v]

The Department of Natural Resources and the Department of Education partnered to deliver the Save it Forward program in 2010. This program had students submit proposals that promoted conservation and EE. The program reached over 8,400 students in over 30 schools and provided over $76,000 in funding.

In 2011, the provincial government launched its EE strategy entitled Moving Forward: Energy Efficiency Action Plan. This plan provides a comprehensive vision for energy consumption, EE, and eventual new directions. It consists of three major goals, notably: (1) supporting a major shift in the uptake of EE; (2) demonstrating provincial government leadership on EE; and (3) advancing actions on EE by collaborating with

---

[iv] The Model National Energy Code of Canada for Buildings (MNECB) was introduced by the National Research Council in 1997 to establish construction standards for building components and features that affect EE in buildings. The MNECB factors in a number of regional differences including construction costs, climate, fuel types and costs.

[v] In Canada, LEED is administered through the Canada Green Building Council and is a market-based rating system that provides third-party verification of green buildings. LEED has become the most recognized green building rating system in North America with thousands of projects registered with and certified by the program.

other jurisdictions.

The OCC launched the Turn Back the Tide public awareness campaign in 2012, an effort to help residents better understand the effects of climate change and impacts of EE. This integrated campaign includes television, print and online components that provide information and tips on ways to reduce GHG emissions, promote EE, and prepare for the impacts of climate change.

The OCC also launched the Market Transformation Framework in 2015. This framework outlines the government's vision for transitioning the buildings, transportation, as well as products and services sectors toward more energy efficient practices with lesser climate change impacts. This framework has been implemented under six key principles:

1. Recognize that market transformation is a long-term process that requires sustained commitment and communication with all interested parties to allow external stakeholders time to prepare for any upcoming changes.

2. Determine what measures are needed in the short and medium term to make progress, but remain flexible enough to adjust to new information, emerging needs, or gaps.

3. Recognize that complementary measures might need to be deployed simultaneously or sequentially.

4. Collaborate with other governments to add value or generate momentum.

5. Promote broad engagement and strong partnerships. Efforts to transform markets are most effective where organizations are aligned, coordinated, and share a common goal. The provincial government seeks opportunities to engage and partner with internal and external stakeholders to effect lasting change in the market.

6. Lead by example through how the government operates. Government actions can result in significant energy and cost savings and have significant influence on the market through the sheer scale of its activities.

Also in 2015, the provincial government released the Greening Government Action Plan. The goal of this five-year plan is to create a culture of environmental sustainability in government operations.

The plan outlines 46 actions across five strategic objectives such as improving the sustainability of new and existing government buildings through EE measures. Other examples of EE measures included in this policy are:

- Use lifecycle costing models to maximize cost-effective energy savings when planning new construction projects or completing energy retrofits.

- Implement EE retrofits in a target number of buildings each year.

- Apply an energy conservation and efficiency lens to all infrastructure renewal decisions and in the selection of products and processes.

- Expand the use of energy management tools in existing buildings.

- Offer a toolkit or Webinar to building managers on best energy management practices.

- Participate in benchmarking initiatives by national organizations, such as the National Executive Forum on Public Property or the Building Technology Transfer Forum, to identify efficiency opportunities.

LEGAL AND REGULATORY FRAMEWORK

Government Code Requirements

The National Building Code of Canada (NBCC) has been adopted in Newfoundland and Labrador and all new homes and buildings must comply with requirements therein. In 2012, significant updates were made to the NBCC to incorporate new EE requirements. The OCC developed the Guide to Building Energy Efficient Homes and Small Buildings to provide a user-friendly overview of the new NBCC requirements which apply to individuals, municipalities, designers, engineers, and contractors active in the building sector of the province. The guide also includes case studies, technical tips, and checklists.

Utility Regulation

The Newfoundland and Labrador Board of Commissioners of Public Utilities (PUB) is the oversight agency that regulates the Utilities.

In 1997, the PUB required the evaluation of rate impacts and cost effectiveness of Newfoundland Power customer conservation programs.[vi] This evaluation requirement was modified in 2016 to allow Newfoundland Power to calculate the cost effectiveness of customer conservation programs using the Total Resource Cost test and the Program Administrator Cost test, while the rate impact evaluation requirement was abolished.[vii]

Program participants must provide certain information in program rebate applications. This information includes technical data such as the R-value of installed insulation, the type of heating in the home, and geographic location. Analyzing these data allows the utilities to accurately estimate the energy savings of each program and perform industry standard economic cost-benefit analyses.

Utility Program Evaluation

The energy savings, market impacts and delivery process effectiveness of Utility customer energy conservation programs are evaluated annually. Additional reviews by third-party evaluators are also conducted. The findings are used to constantly refine program design and implementation and support further planning.

Table 6-2 summarizes the primary methods used by the Utilities to plan and evaluate customer conservation initiatives.

FINANCIAL MECHANISMS

Utility Incentive Programs

The Utilities have offered programs on a joint and coordinated basis since 2009. Table 6-3 presents a list of programs offered since the implementation of the takeCHARGE partnership.

[vi]   As described in Newfoundland Power Inc.—2009 Conservation Cost Deferral Application, Section 2: Proposed Customer Program Portfolio filed with the Board October 29, 2008. The Total Resource Cost test measures net program benefits in terms of utility system avoided costs against utility and customer costs for the program. The Ratepayer Impact Measure test measures the impact on customer rates as a result of changes in utility revenues and operating costs due to utility conservation programs.

[vii] The Program Administrator Cost test measures net costs of utility conservation programs as a resource option based on the costs incurred by the program administrator, including incentive costs and excluding any net costs incurred by participants. This change was approved in Order No. P.U. 18 (2016).

Table 6-2: Conservation Planning and Evaluation Tools

Evaluation and Planning Tool	Frequency	Purpose
Conservation Potential Study	5 years	Identify opportunities for energy savings
5-Year Plan	3-5 years	Determine programs and supporting activities
External Review	2 years for each program	Program process effectiveness and market transformation impacts
Post-Implementation Review	6-12 months after new program launch	Program process effectiveness
Process Evaluation	Annual	Program process effectiveness
Economic Testing	Annual	Economic and energy savings impacts
Customer Survey	Annual	Assess customer awareness and responsiveness
Verification Audits	Ongoing	Gather customer feedback on programs and customer program compliance
Partner Consultation	Ongoing	Gather feedback and assess customer responsiveness

Table 6-3: Five-Year Energy Conservation Plan Programs by Year

Five-Year Energy Conservation Plan	2008-2013	2012-2016	2016-2020
Residential ENERGY STAR® Windows Incentive Program	X	X	
Residential Programmable and Electronic Thermostats Program	X	X	X
Residential Basement and Attic Insulation Program	X	X	X
Commercial Lighting Program	X	X	
Industrial Energy Efficiency Program	X	X	X
Isolated Systems Community Program		X	X
Block Heater Timers Program		X	
Isolated Systems Business Efficiency Program		X	X
Instant Rebates Program		X	X
Appliance and Electronics Program		X	X
Heat Recovery Ventilator Rebate Program		X	X
Business Efficiency Program		X	X
Benchmarking Program			X

The following presents a brief description of each program currently offered by the Utilities, as outlined in the Five-Year Conservation Plan: 2016-2020.

Residential Basement and Attic Insulation Program

The objective of this program is to provide incentives to increase the insulation R-value in residential basements, crawl spaces, and attics, thereby augmenting the efficiency of home building envelopes. Eligibility is determined based on annual energy usage and is limited to electrically heated households undergoing home retrofit projects. Customers receive a 75% rebate on the costs of basement wall and ceiling insulation material and a 50% rebate on attic insulation material costs. Both rebate components reimburse up to a maximum of $1,000 in expenses.

Residential Programmable and Electronic Thermostats Program

This program encourages the installation of programmable and electronic thermostats so customers can better control home temperatures and save energy. These high performance thermostats allow customers to lower temperature settings at night or when away. Eligibility is determined based on annual energy usage and is limited to electrically heated homes. Home retrofit projects and new home constructions are eligible. The program offers incentives of $10 for each programmable thermostat and $5 for each electronic high-performance thermostat.

Heat Recovery Ventilator Rebate Program

This program encourages customers to purchase high-efficiency heat recovery ventilators (HRVs) to improve the efficiency of their home. Eligible measures include HRV models that have a sensible recovery efficiency of 70% or more. Customers who purchase high-efficiency HRVs qualify for a rebate of $175. All customers are eligible, regardless of the age of the home or heat source.

Benchmarking Program

This program encourages customers to modify their energy consumption behaviors. Participants receive home energy reports that provide insight into the electricity use of their homes. The reports also help participants understand changes in their usage over time and compare usage to similar homes. The reports also contain practical tips on how to save energy moving forward. The program has an online component that allows customers to engage even further through weekly challenges and personalized saving plans.

Instant Rebates Program

This program promotes a variety of smaller technologies such as LED lighting and smart power bars through instant rebates available at the cash register of participating retailers. All customers are eligible.

Appliance and Electronics Program

This program encourages customers to purchase high-efficiency appliances. Participants receive incentives of $100 for select energy efficient washers and freezers, and $30 for eligible televisions. All customers are eligible.

Business Efficiency Program

This program aims to improve electrical EE in a variety of commercial facilities. Program components include financial incentives based on energy savings, as well as additional financial and educational support to enable commercial facility owners to identify EE potential and carry out demand-reduction projects.

This program is available to existing commercial facilities that demonstrate they can save energy or reduce energy demand by installing more efficient equipment and systems. Custom project incentives and rebates for specific measures on a per unit basis are offered under this program.

Isolated Systems Business Efficiency Program

This program targets commercial customers located in remote communities serviced by NFL Hydro's isolated systems. This custom program offers incentives based on the energy savings from efficiency improvement projects. This allows customers to implement energy efficient technologies that suit their specific buildings, equipment, and operations.

Isolated Systems Community Program

This program includes direct installations of energy efficient products at no cost to homes and businesses. The program also focuses on customer education and building capacity in communities by hiring and training local representatives. These representatives work in their own communities to promote the program, provide information on energy use, and install products.

Industrial Custom Projects Program

This is a custom program that responds to the unique needs of transmission-level industrial customers. It provides financial support for project implementation costs and engineering feasibility studies.

Utility Educational Programs

Aside from offering financial incentives, the Utilities provide customer education and conservation awareness services primarily through customer outreach activities. These activities involve mass media marketing, community outreach such as school programming, and trade ally development and partnerships.

These educational initiatives have recently been expanded to include a program focused on promoting mini split heat pumps. Components include education and marketing initiatives that target customers, as well as direct engagement with certified installers and suppliers.

In 2013, the Utilities launched the Kids in Charge, or K-I-C Start school program which provides EE and conservation education support to students throughout Newfoundland and Labrador. Activities include in-class presentations and holding an annual contest for elementary and high school students. In 2014, takeCHARGE partnered with the OCC to extend this program to the Hotshots pilot program. Through the HotShots pilot, the province provided funding and support for additional in-class presentations, curriculum teaching materials, and a contest for high school students.

Utility On-bill Financing

The Utilities offer on-bill financing on a number of energy efficient technologies.[viii] Financing is available to residential homeowners with an active electricity account. Financing is available upon credit approval at a rate of prime plus 4%.

Table 6-4 lists the energy efficient products that qualify under the program, along with the maximum financing amount and term.

Table 6-4: Utility On-Bill Financing Technology, Amount, and Term

Product	Maximum Amount Financed	Maximum Term of Financing
R-2000 Upgrades	$10,000	60 Months
Electric Heating Systems & Heat Pumps	$10,000	60 Months
Heat Recovery Ventilation Systems (HRVs)	$5,000	60 Months
Electronic & Programmable Thermostats	N/A	36 Months
Basement and Attic Insulation	$5,000	60 Months

Government Programs

Recently, the provincial government announced $9 million in funding over three years for two new home EE programs. The government wants to provide financial assistance to homeowners looking to reduce energy costs by improving the EE of their homes.

More specifically, the Home Energy Savings Program (HESP) replaced the Residential Energy Efficiency Program in July 2017. It

---

[viii] Newfoundland and Labrador Hydro offers financing on heat pumps only.

provides low-income households with grants of up to $5,000 for cost-effective upgrades in electrically heated existing homes with focus on insulation and air sealing. The HESP is delivered by the Newfoundland and Labrador Housing Corporation.

The second new program, the Home Energy Efficiency Loan Program (HEELP) provides low-interest loans for EE home upgrades. Through the HEELP, qualifying homeowners whose homes are electrically heated can receive financing of up to $10,000 over five years for heat pumps and insulation. The program offers low-interest financing at prime plus 1.5% and is delivered by the Office of Climate Change in partnership with the Utilities. The program launched in October 2017.

Non-profit Programs

The CCNL offers three main youth employment programs: Green Teams; Environmental and Cultural Hiring Opportunity (ECHO); and Internships. Green Teams targets youth teams ages 16 to 30, while ECHO concentrates on individual students and Internships engages recent graduates. These programs strive to create close connections with community partners.

CCNL offers other environmental and cultural conservation initiatives. The Climate Change Education Centre, supported by the Department of Municipal Affairs and Environment, delivers in-school education presentations on climate change and the impacts climate change has in the province.

Chapter 7

Northwest Territories

ENERGY SECTOR

The Northwest Territories (NWT) is a territory of Canada with an area of 1.346 million km². According to the latest statistics from 2016, the population of the NWT comprises 44,469 people living in 33 communities.[1] The NWT economy relies heavily on resource industries such as mining, and communities are widely dispersed. These factors have major impacts on energy consumption distribution by sector, as illustrated in Figure 7-1.

Throughout the 1999-2000 to 2013-2015 period, energy consumption in the NWT increased by 22% to reach 452 ktoe. Most of that increase was due to the industrial sector whose energy consumption nearly doubled and reached 55% of total energy consumption for the

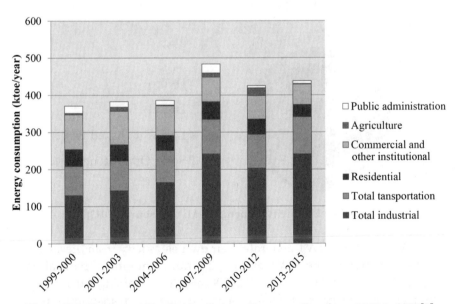

Figure 7-1: Northwest Territories Energy Consumption from 1999 to 2015[2]

2013-2015 period. The energy consumption of the transportation sector also increased faster than total consumption, by 27%, and represented the second most energy intensive sector with 22% of total consumption, while the combined energy consumption of the industrial and transportation sectors represented 78% of total consumption in 2013-15.

Conversely, the energy consumption of the commercial and institutional sectors combined declined by 41% while the residential sector declined by 28% between the 1999-2000 and 2013-14 periods. Their combined relative importance to total energy consumption is nearly 20% for the period. The public administration and agriculture sectors constitute the remaining 2% of energy consumption.

The NWT has great hydroelectric resources which represent close to 75% of the electricity generation mix (Figure 7-2). The eight communities located in the Great Slave Lake region (Southern NWT) are the most populated and supply electricity from hydropower. The communities of Inuvik and Norman Wells supply electricity from liquefied natural gas (LNG), while the 23 other communities rely on diesel-fired power plants. Overall, diesel-fired power represents approximately one fifth of the total electricity generation mix, while LNG, purchased power and renewables represent less than 5%.

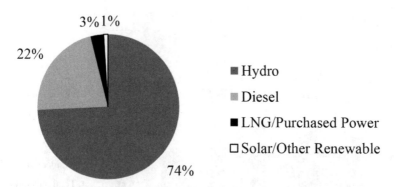

Figure 7-2: Electricity Generation Mix by Source, 2013[3]

Mines in the NWT generate their own electricity from diesel generators. Although the NWT extracted a huge amount of natural gas (83 million m^3, 2016) and oil (522,000 m^3, 2016), much of the energy sources used for generating electricity, heating and transportation are imported.[1]

ENERGY EFFICIENCY BACKGROUND

Due to the NWT geographic and economic situation, the development of public infrastructure such as roads and transmission lines is challenging, which exerts great pressure on the cost of energy production, transmission and distribution infrastructures. Additionally, the long cold winters almost double the energy use per capita (428 GJ, 2009/2010) compared to the Canadian average (227 GJ, 2009/2010).[4] These factors affect the capability of delivering affordable and reliable energy, consequently increasing the cost of living and doing business.

The NWT has high levels of imported oil, energy costs, energy usage per capita and GHG emissions per capita.

Therefore, the main drivers for EE and renewable energy are high energy costs and environmental stewardship. Indeed, supporting the use of EE is one of the most effective ways to reduce energy use and costs and mitigate the impacts of climate change. Figure 7-3 outlines the main NWT EE initiatives in chronological order.

The establishment of Arctic Energy Alliance (AEA) in 1997 marked the beginning of sustainable development in the NWT. This non-profit organization was founded to help communities, consumers, producers, regulators and policymakers work together to reduce the costs and environmental impacts of energy and utility services. However, the Canada-NWT Agreement on the Transfer of Federal Gas Tax Fund (GTF) established the actual development of EE. The Government of the Northwest Territories (GNWT), represented by the Department of Municipal and Community Affairs (MACA), and the federal government jointly implemented the Control Management Framework (CMF) of the GTF. It arranged the formal governance structure, delivery mechanism and risks of the program. The GTF provided funding to community governments to develop their municipal services in water supply, wastewater, solid waste, community energy systems, active transportation and capacity building. Funding was delivered to eligible recipients in two payments per year. The Agreement on the Transfer of Federal GTF required communities to submit a five-year Capital Investment Plan (CIP) starting from March 2007 and an Integrated Community Sustainability Plan (ICSP) from March 2010. Although some communities did not sign an agreement under the GTF, the MACA required communities to nonetheless provide the necessary documentation. The MACA also required community governments to determine their long-term sustain-

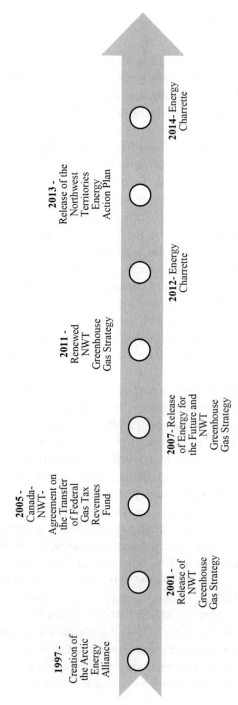

Figure 7-3: Timeline of Main Energy Efficiency Events in NWT

ability plans and objectives and complete a Community Expenditure Report (CER) on a quarterly basis starting from 2009. Communities also needed to submit audited financial statements each year.[5] As of 2006, the MACA has been providing funding to AEA to support community governments in completing their energy plans. In 2009, the AEA also increased its presence throughout NWT with the opening of five regional offices outside Yellowknife.

In 2007, the GNWT committed to reducing GHG emissions and the cost of energy by releasing the Energy for the Future Plan. The GNWT also released the complementary 2011 NWT Greenhouse Gas Strategy update, whose initial version dated back to 2001 in which the government had committed to adopting the national target of reaching 6% below 1990 emission levels by the year 2012. The focus of the NWT GHG Strategy 2001 was to:

- "Identify and coordinate northern actions to begin to control GHG emissions; and

- Assist in developing and contributing a northern perspective as part of Canada's national climate change implementation strategy."[6]

The strategy contained 20 measures focused on five areas. In 2005, the Department of Environment and Natural Resources began reviewing how the strategy was progressing. The review revealed that new actions had already been taken which were not included in the 2001 GHG Strategy.

A new version of the strategy was released in 2007, which contained:

- "The same goals, objectives and principles from the 2001 version;

- An update of ongoing actions and new measures that will be initiated;

- Better ways to track, report and coordinate actions taken to control GHG emissions from the NWT; and

- New short and long-term targets for the NWT."[6]

Consultations with all stakeholders ensued in 2011, resulting in the release of another strategy for the 2011-2015 period. This latest

GHG strategy established 34 actions in all sectors. The main goal was to achieve 2005 GHG emission levels (1,500 kt) by 2030.[7]

In 2012, an energy charrette was held, which gathered over 120 representatives from communities, governments, non-government organizations, industry, and energy experts. The charrette led to the release of the 2013 NWT Energy Action Plan in December of said year. The three-year action plan included initiatives in five different areas and further increased AEA funding by nearly five times more for 2014-15 than for 2006-07.

Another energy charrette was held in 2014 to more greatly focus on necessary territorial government actions in the energy sector. The AEA allocated budget for 2015-2016 was $3,142,500. The four most popular programs for 2015/2016 were: Energy Efficiency Incentive Program (EEIP), Commercial Energy Conservation and Efficiency Program (CE-CEP), Alternative Energy Technologies Program (AETP), and Community Renewable Energy Program (CREP).[7] Figure 7-4 illustrates AEA funding increases over the 2005 to 2016 12-year period.

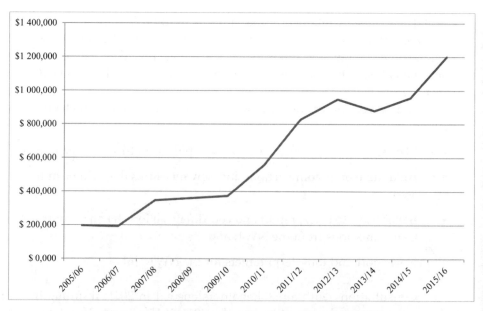

Figure 7-4: AEA Core Funding from 2005 to 2016[8]

STAKEHOLDERS

Institutions in Charge of Strategy[9], [10]
    The field of EE has been mainly influenced by the GNWT via its departments and agencies. Energy issues fall under the purview of the following governmental departments:

- Department of Industry, Tourism and Investment;

- Department of Environment and Natural Resources;

- Department of Finance;

- Department of Municipal and Community Affairs (MACA);

- Department of Infrastructure (Public Works and Services [PWS]).

The GNWT has established the Ministerial Energy Coordinating Committee (MECC), which helps coordinate the work between multiple departments, identify issues, track efforts and support decision making. It is chaired by the Minister of Industry, Tourism and Investment whose department is responsible for energy planning and policy development. The MECC is supported by the Deputy Ministers Energy Coordinating Committee. In 2015, the Department of Finance issued the Territorial Power Subsidy Policy under which the Territorial Power Subsidy Program is managed.[11] The Territorial Power Subsidy Program ensures affordable electricity for all residential users throughout the NWT. The program subsidizes residential consumers living outside Yellowknife with an amount equal to the difference between their electricity rate zone and the Yellowknife reference rate for electricity consumption. The Department of Environment and Natural Resources is in charge of energy DSM, EE and conservation programs managed through the AEA. It "works to promote and support the sustainable use and development of natural resources and to protect, conserve and enhance the Northwest Territories (NWT) environment for the social and economic benefit of all residents."[12] The MACA facilitates sustainable development by community governments, while the Department of Infrastructure is responsible "for the strategic planning of long term investment in the development of public transportation infrastructure and energy production, and distribution systems for the NWT."[13] The latter is also responsible for the publication of energy plans, energy conservation reports, energy reports and other plans (biomass plans, alternative energy plans, etc.).

The NWT Housing Corporation is the agency through which the GNWT achieves public housing EE and conservation goals.

Institutions in Charge of EE Programs
Arctic Energy Alliance (AEA)
Almost all EE programs are delivered by the AEA, the lead GNWT agency in the EE sector. The AEA is a not-for-profit organization whose mandate is to help communities, consumers, producers, regulators and policymakers reduce both the cost and environmental impacts of energy and utility services in the territory.[14] In addition to the AEA main office in the capital of Yellowknife, it maintains five regional offices in different parts of the territory.

Northwest Territories Power Corporation (NTPC)
The AEA works in collaboration with NTPC on many EE projects. Since NTPC is in charge of the generation and transmission of electricity as well as hydro development in the territory, it is the most important utility in the NWT. It is a vertically integrated electricity monopoly and subsidiary of NT Hydro which is government owned. NTPC is regulated by the NWT Public Utilities Board (PUB). As the main electricity distributor, NTPC plays a large role in the promotion of EE, especially through the multiple energy conservation tools available on its website. It also promotes funds managed through AEA programs.

POLICIES AND STRATEGIES

The most recent NWT Energy Action Plan was issued by the GNWT at the end of 2013. This plan presented the actions to be pursued from 2014 to 2016. The programs identified in this plan are almost all still offered. The Action Plan was the result of the energy charrette held in 2012 to hear community concerns on the energy situation. Another energy charrette was held in 2014 after the release of the plan, which resulted in short-term action focusing more on EE and conservation.[15]
The Energy Action Plan was aligned with the NWT Power System Plan (PSP) developed by NT Energy Corporation. The NWT PSP presents the electricity infrastructure objectives until the 2030s. The previous version of the Energy Action Plan was released in 2007, which led to the publication of the Greenhouse Gas Emission Strategy in that same

year. The action plan was updated in 2011.

The recent NWT Energy Action Plan recommends 33 actions in four areas:[16]

- Energy conservation and efficiency;
- Energy supply;
- GNWT leadership;
- Policy and planning.

The plan proposes the possibility of establishing a NWT Energy Efficiency Act, which was reinforced in the GNWT Response to the 2014 NWT Energy Charrette Report. As of February 2018, no EE act or legislation had yet been adopted by the NWT.

The three territorial governments (NWT, Yukon and Nunavut) have developed collaborative documents collectively entitled A Northern Vision. These documents outline the common challenges, opportunities and issues affecting their territories, including EE and conservation initiatives, in a document entitled Paths to a Renewable North published in 2011.[17]

Carbon

Officials in NWT are still discussing carbon pricing which needs to be introduced throughout all of Canada by 2018.[18] The residents of NWT will have the opportunity to discuss approaches for a carbon pricing mechanism through the Climate Change Strategic Framework.[19] The NWT is concerned about how such a mechanism will be applied to the small northern economies which mostly rely on diesel generators. The money levied by the carbon tax has also been discussed, notably how to invest in new renewable technologies and resolve other issues.

The best solution for the GNWT would perhaps be to develop a carbon tax mechanism rather than a cap-and-trade system because the GNWT already has a taxation system on many fossil fuels.[20]

EE in the Public Sector

The government has taken many steps to increase EE in the public sector. The Capital Asset Retrofit Fund (CARF) provides funding for EE upgrades on government buildings. The fund is financed through savings from previous energy improvements.[21] In 2009, municipal building retrofits in Fort Smith reduced GHG emissions by approximately 860 tons of CO_2e.[24] In 2016, the CARF was used mainly to encourage the installation of biomass boilers.

As of February 2018, there are two available AEA programs for EE in public buildings. The Community Government Building Energy Retrofit Program supports local governments in identifying the current energy use of public buildings, the most suitable measures to be developed and the cost of implementing measures, while the Alternative Energy Technologies Program provides funds to implement renewable energy projects in governmental buildings.[22]

Furthermore, all new GNWT buildings must be designed to be 10% more energy efficient than a facility built in compliance with the NECB 2011. Currently, PWS is reviewing the new National Energy Code for Buildings 2015 to determine new performance targets.

Good Building Practice for Northern Facilities

The GNWT released the Good Building Practice for Northern Facilities guide. The last edition was released in 2011 and updated in 2013. This guide is a mandatory standard building practice for new GNWT constructions no matter the size. It also provides a clear design and building method for any Northwest Territorian wanting to build a more sustainable and efficient building in the territory.[23]

In addition to the Good Building Practice for Northern Facilities, PWS has another initiative in the public building sector. PWS monitors the budgets of all government premises to assess electricity consumption and estimate GHG emission reductions. The collected data is used to benchmark energy consumption over time and guide the implementation of EE measures.

A Greenhouse Gas Strategy 2011-2015[25]

The most recent version of the Greenhouse Gas Strategy was released in 2011. The first of three former editions was published in 2001 and revised in 2005 and 2007. Importantly, the effects of climate change are more tangible in the NWT than the rest of the world:

- The annual average temperature in the NWT for the last 50 years increased by 2 to 2.7 °C, while the global average has increased by 0.5°C;[26]

- Arctic sea ice thickness has decreased by 1.7 m, from 3.6 m to 1.9 m, from 1980 to 2008 (nearly a 47% decrease). According to NR-Can, even under moderate warming scenarios, 40 to 75% of the building foundations in the town of Inuvik could be damaged if no action is taken.

The GHG emissions of the NWT constitute a very small portion of world emissions, but the GNWT wants to demonstrate its commitment to this issue by reducing the future impacts of warming temperatures. The main targets of the government's Greenhouse Gas Strategy 2011-2015 were to:

- "Stabilize emissions at 2005 levels (1,500 kt of CO_2e) by 2015.

- Limit emission increases to 66% above 2005 levels (2,500 kt of CO_2e) by 2020.

- Return emissions to 2005 levels (1,500 kt of CO_2e) by 2030."

In 2014, NWT emissions were estimated at 1,531 kt of CO_2e.[27]

Energy Priorities Framework (2008)[28]
The Energy Priorities Framework was released in 2008 and identifies five key action areas for which key energy priorities and initiatives have been identified:
- Energy conservation and efficiency;
- Alternative energy and emerging technologies;
- Energy policy and planning;
- Energy development and supply;
- Reducing GNWT energy use.

LEGAL AND REGULATORY FRAMEWORK

As previously mentioned, the NWT has adopted no EE act or legislation. The most significant change in NWT energy policy is in electricity rates. Between 2008 and 2010, many discussions were held and actions taken, such as the establishment of seven rate zones, the introduction of riders and direct accountability, improving dialog between the utility and customers, etc. In March 2015, the Territorial Power Subsidy Policy was released by the Department of Finance. It states: "Residential electricity consumers in each community will be subsidized an amount equal to the difference between the residential electricity rate for their electricity rate zone and the Yellowknife reference rate for electricity consumption up to:

(a)   1,000 kilowatt hours per billing for a maximum of seven billing periods annually, and

(b)   600 kilowatt hours per billing for a maximum of five billing periods annually."[29]

Some AEA programs are designed to promote EE, namely:

• Core Funding for Regional Projects, Coordinators and Offices – funding support for AEA core and substantive activities (trade-shows, community events, fares, competitions, and others);

• Energy Management Program – to fund energy management and communications (promotion of EE through the AEA website and Facebook page, Yardstick Audits, Fuel Cost Library and Technical Committee, training events, conferences, case studies, and others);

• Energy Rating Service – for homes under license agreement with NRCan. The program provides a sustainability evaluation of residential buildings using a rating of 0 to 100 (air leakage, energy consumption). AEA offers advice to homeowners based on assessment results;

• Electric Vehicle Demonstration Project – in 2015, AEA tested an electric vehicle in Yellowknife. AEA used a data logger to record and monitor the performance of the vehicle in northern conditions.

EE in Building Codes

The NWT enforces local energy requirements for all buildings constructed using government funds that meet NECB 2011 requirements. [30] A new version of the code was released in 2015, and the GNWT is currently reviewing program targets.

FINANCIAL MECHANISMS

Figure 7-5 provides a breakdown of GNWT funding for EE and conservation. The government planned to spend nearly $9 million in 2016-2017. Approximately 60% of these investments are in EE and conservation programs.

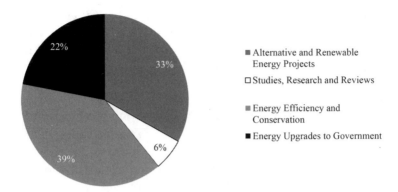

Figure 7-5: GNWT Energy Funds for 2016/2017[31]

According to the AEA report for 2015/2016, the agency had an annual budget of $3,142,500, with additional one-time supplementary funding of $840,000 for 2015/2016 targets and $760,000 for 2016/2017 EE targets. The AEA provided $598,802 in rebates (excluding EEIP rebates).[32] The following programs are offered by the AEA:

- Energy Efficiency Incentive Program (EEIP);

- Alternative Energy Technologies Program (AETP);

- Community Building Energy Retrofit Program (CBERP);

- Commercial Energy Conservation and Efficiency Program (CECEP).

EEIP provides rebates to homeowners, businesses and non-profit organizations for the purchase of efficient equipment, while CECEP provides rebates for energy upgrades in commercial buildings. AETP and EEIP provide rebates for renewable energy sources to residents, businesses as well as communities. Table 7-1 outlines a breakdown of the budgets allocated by the GNWT to the four programs.

The results (excluding EEIP) of the 2015/2016 fiscal year are provided in Table 7-2. As of the end of March 2016, a budget of $170,298.74 was spent on 420 rebates through EEIP.

Also, PWS operates an initiative in the public building sector – the Capital Asset Retrofit Fund (CARF) program. This program was created in 2009 to help the GNWT reduce GHG emissions, electricity bills and energy consumption in public buildings.

Discussions are being held on changes to the Cities, Towns and Villages Act to improve incentive programs. These changes need to offer property and business owners access to long-term and low-cost municipal financing for the purchase of efficient technologies.[33]

Table 7-1: Breakdown of 2015/2016 Budgets for AEA Programs

	2015/2016 Budget:
Energy Efficiency Incentive Program (EEIP)	$300,000
Alternative Energy Technologies Program (AETP)	$410,000 plus an additional $150,000
Community Building Energy Retrofit Program (CBERP)	$200,000
Commercial Energy Conservation and Efficiency Program (CECEP)	$200,000 plus an additional $50,000

Table 7-2: 2015/2016 Results

Metric	Total (excluding EEIP)
Total number of rebates	72
Total rebate	$598,802
Average rebate	$8,317
Total capital cost	$3,468,957
Estimated annual savings	$425,492
Estimated annual electricity savings (MWh/year)	56.2
Estimated annual oil savings (liters)	122,014
Estimated annual CO_2e savings (t)	375

Chapter 8

Nova Scotia

Mr. Stephen MACDONALD, EfficiencyOne

ENERGY SECTOR

After increasing by 10% from 1995 to 2006, Nova Scotia's energy consumption reached 3,406 ktoe in the 2013-2015 period, a 10% reduction compared to 1995-97 levels. The energy consumption of almost every sector also decreased over the 20-year period from 1995 to 2015, except for the residential sector which consumed the same amount of energy in 1995-97 as in 2013-15. Throughout the latter period, the residential sector consumed 24% of total provincial energy usage. The transportation sector consumed the most energy, accounting for 43% of total energy consumption in 2013-15. The commercial and industrial sectors accounted for 13% and 14% respectively during the same period, though these sectors each decreased consumption by 20% compared to 1995-97 levels. The agriculture and public administration sectors consumed 1% and 5% respectively in 2013-15.

Figure 8-1 illustrates average energy consumption in three-year periods from 1995 to 2015.[1]

ENERGY EFFICIENCY BACKGROUND

Nova Scotia has continuously pursued electricity DSM since 2008. Over this period, three different administrative models have been used to plan and carry out DSM activities. Chronologically, these models are:

- Utility administration of DSM by Nova Scotia Power Incorporated (NS Power, 2008-2010);

- Independent administration by Efficiency Nova Scotia Corporation (ENSC) under the Efficiency Nova Scotia Corporation Act (2010-2014);[2] and

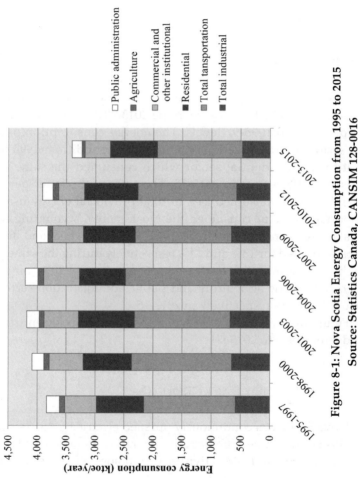

Figure 8-1: Nova Scotia Energy Consumption from 1995 to 2015
Source: Statistics Canada, CANSIM 128-0016

- Independent administration of the Efficiency Nova Scotia (ENS) franchise by EfficiencyOne under the Public Utilities Act (2015-current).[3]

Beginning in 2008, NS Power began administering DSM programs with the purpose of reducing electricity usage in Nova Scotian homes and businesses. In April 2008, Dr. David Wheeler released a report entitled Stakeholder Consultation Process for an Administrative Model for DSM Delivery in Nova Scotia (Wheeler Report).[4] This report outlined strong stakeholder preference for an independent DSM delivery model over other possible options and influenced subsequent policy direction for DSM activities in Nova Scotia. Stakeholders were uncomfortable with the apparent conflict of interest of allowing a regulated monopoly, NS Power, to sell electricity while also being the administrator of EE programs whose objective is to help customers reduce electricity consumption.

In October 2009, the Nova Scotia Legislature passed the Efficiency Nova Scotia Corporation Act (the ENSC Act) which came into force in January 2010. The ENSC Act established ENSC as the independent administrator of DSM activities in the province effective in 2010. The ENSC Act also provided statutory authority for the key features of the independent administrator model, including the supervisory role of the Nova Scotia Utility and Review Board (UARB) for DSM planning purposes and overall regulatory oversight.

The ENSC model operated unchanged until 2014 when the NS Legislature passed the Electricity Efficiency and Conservation Restructuring (2014) Act.[5] This Act, by virtue of amending the Public Utilities Act, influenced the model for DSM in Nova Scotia in several important ways, the most important of which are listed below:

- ENSC ceased to exist and no longer delivered DSM activities as of January 1, 2015. The Efficiency Nova Scotia (ENS) franchise was created to replace it. The ENS franchise is owned by the Province of Nova Scotia and administrative duties are awarded by the Minister of Energy. The ENS franchise "gives the franchise holder the exclusive right to supply Nova Scotia Power Incorporated with reasonably available, cost-effective electricity efficiency and conservation activities."[6]

- EfficiencyOne was created as a non-profit corporation and was awarded the ENS franchise for a 10-year term from January 1, 2016 to December 31, 2025, in addition to an initial one-year transition period for 2015.[7]

- The supply of electricity efficiency and conservation activities would now be administered by the ENS franchise on a contractual basis with NS Power, and the supply agreement was approved by the UARB. For the 2015 transition year, NS Power and ENS were deemed to have entered into an initial supply agreement.[8]

- The legislation states that the UARB "shall establish such performance requirements for the franchise holder as it considers appropriate."[9]

- Sections of the ENSC Act that allowed for a rate-rider to be collected on NS Power bills were repealed.[10]

- The new model treats EE as a system resource for NS Power, and recovery of NS Power's investment in DSM is now embedded in its general electricity rates. As such, ENS becomes Canada's first electricity efficiency utility.

Several important aspects of the previous model were preserved in the revised version, including independent evaluation and verification processes, stakeholder representation and ongoing consultation through the DSM Advisory Group, and the establishment of specific targets for annual energy and system-coincident peak demand savings.

ENS offers programs with financial incentives and support to help residential, business, non-profit and institutional participants save energy via two branches:

1. Electricity efficiency and conservation services provided by the ENS franchise under a supply agreement with NS Power and regulated by the UARB; and

2. Energy saving services carried out pursuant to a contract with the Province of Nova Scotia (PNS contract or PNS energy savings). The Province identifies policy priority investment areas; the current PNS contract is structured to help low-income Nova Scotian homeowners and renters save energy. While this work is performed outside the ENS franchise, it is carried out under the ENS brand in accordance with the Franchise Agreement.

ENS residential and commercial initiatives are continuously evolving based on factors that include evaluation and verification recommendations, industry knowledge, best practices, market needs, stakeholder and participant feedback, as well as available resources. As such, ENS residential and commercial initiatives have evolved significantly since Nova Scotia began pursuing electricity efficiency in 2008.

This focus on continuous improvement and innovation ensures that ENS programs and services remain industry-leading and highly responsive to participant needs. ENS's proven program management capabilities are complemented by solid expertise in research, regulatory affairs, marketing, outreach and education, and stakeholder engagement.

ENS currently offers six core EE programs, each with its own program component(s), as outlined in Table 8-1. The main initiatives pursued by these programs and their components are detailed further below.

Table 8-1: ENS Energy Efficiency Programs

PROGRAM	PROGRAM COMPONENT
Residential	
Efficient Product Rebates	Instant Savings
	Appliance Retirement
Existing Residential	Home Energy Assessment
	Green Heat
	Efficient Product Installation
New Residential	New Home Construction
Business, Non-Profit and Institutional (BNI)	
Efficient Product Rebates	Business Energy Rebates
Custom Incentives	Custom
	Energy Management Information Systems
	Strategic Energy Management
Direct Installation	Small Business Energy Solutions

The Instant Savings component of Residential Efficient Product Rebates generates net energy savings by providing customers with financial savings on energy-efficient product purchases, while Appliance Retirement incents participants to retire old inefficient appliances.

The Home Energy Assessment component of Existing Residential incents participants to have a complete analysis conducted of their home's EE and pursue EE opportunities.

Green Heat participants are incented to upgrade their home heating equipment, while Efficient Product Installation provides no-cost home installations of energy efficient products in houses and apartments. The New Home Construction component of New Residential incents participants to build energy efficient housing by providing custom recommendations on building plans and incentives to pursue energy efficient options.

For the commercial sector, the BNI Business Energy Rebates component provides instant and mail-in rebates to offset the cost of premium energy efficient products. The Custom component of Custom Incentives helps reduce the energy use of businesses by providing technical assistance and financial support. Moreover, the Custom component includes the Building Optimization service which provides technical support and incentives to optimize the efficiency of existing building systems.

The Energy Management Information Systems component of Custom Incentives provides a monitoring system to analyze where and how businesses use energy, while the Strategic Energy Management component offers funding and support for enterprises to implement long-term energy and cost-saving initiatives.

Over the past five years and further to ENS's supply agreements with NS Power, the ENS Residential and BNI initiatives have achieved significant net incremental electricity savings as outlined in Figure 8-2.

As previously mentioned, ENS also has a PNS contract to deliver energy savings for low-income Nova Scotians. These energy savings are generated through a dedicated low-income program called Home-Warming and several program components listed in Table 8-1, including Efficient Product Installation, Home Energy Assessment and New Home Construction. The HomeWarming program offers no-charge energy assessments and free home upgrades for eligible homeowners with non-electrically heated homes.

Effective 2015, a separate organization called Clean Foundation began administering a new electrical segment of the HomeWarming program for eligible homeowners with electrically heated homes. Funding is sourced from a charitable donation provided by NS Power. Additional PNS energy savings are generated through some of the aforementioned ENS program components and are funded by the Province of Nova Scotia. Figure 8-3 presents the net incremental PNS energy sav-

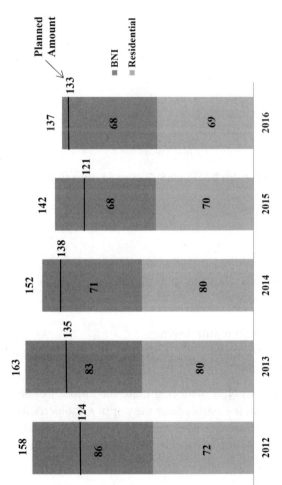

Figure 8-2: 2012-2016 Incremental Annual Electricity Savings (GWh)[i]

[i] Numbers may not sum due to rounding. These electricity savings have avoided the need for NS Power to generate 8.6% more electricity.

ings achieved by ENS for the past four evaluation cycles; energy savings achieved by Clean Foundation are not accounted for in this figure.

Figure 8-3: 2012-2016 PNS Energy Savings (GJ)

STAKEHOLDERS

The following describes the roles and responsibilities of EE stakeholders in Nova Scotia.

Electrical Energy and Demand Savings

- EfficiencyOne prepares DSM resource plans with the advice of expert EE consultants;

- EfficiencyOne and NS Power negotiate the amount of energy savings and financial investment for the supply of electricity efficiency and conservation activities over the contract period (NS Power does not directly influence the design or direction of the DSM resource plan);

- The UARB reviews and approves a final agreement between EfficiencyOne and NS Power; and

- Where NS Power and the franchise holder are unable to reach a final agreement, the UARB mediates an agreement between the parties.

PNS Energy Savings

- EfficiencyOne and the Province of Nova Scotia work together to negotiate a contract and set energy savings targets based on, among other things, past program results;

- The Province of Nova Scotia identifies policy priority investment areas (the current PNS contract is structured to help low-income Nova Scotian homeowners and renters save energy); and

- EfficiencyOne and the Province of Nova Scotia meet regularly to report and track progress toward agreed-upon energy savings targets.

NS Power is a regulated de facto monopoly that provides 95% of electricity generation, transmission, and distribution in the province. [11] The utility serves 500,000 residential, commercial, and industrial customers.[12] Also, several small municipalities operate their own electrical utilities and the UARB is the provincial regulator for all electrical utilities.

Through ENS programming, Nova Scotians have reduced annual electricity consumption by 9% as of the end of 2016. To provide for Nova Scotia's remaining electricity needs, NS Power electricity generation resources include coal, petcoke, natural gas, wind, hydro, tidal, and biomass.[13] See Figure 8-4 for a detailed breakdown of the fuel mix.

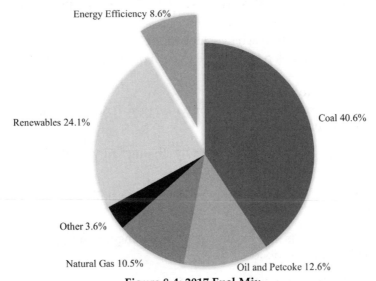

Figure 8-4: 2017 Fuel Mix

POLICIES AND STRATEGIES

As previously mentioned, EfficiencyOne is the independent organization charged with leading Nova Scotia's EE and conservation efforts in terms of long-term strategy and planning as well as ENS franchise day-to-day operations. The UARB regulates and approves savings targets and funding levels put forth by EfficiencyOne.

EE falls under the purview of The Nova Scotia Department of Energy. When preparing the Province's new EE delivery model in 2014, the department reaffirmed its commitment to EE and the leadership role Nova Scotia plays in the field:

> "Energy efficiency has many economic, environmental, and social benefits. Energy efficiency is the cheapest and cleanest way to meet our electricity and energy needs. The economic activity associated with efficiency upgrades and expertise stays in the province, creating good jobs. It helps Nova Scotians and Nova Scotia businesses save money, which makes electricity more affordable and our companies more competitive."[14]

EfficiencyOne recently completed its 2016-2020 strategic plan, an ambitious document whose purpose is "changing lives by unleashing the power of efficiency."[15] The strategy is founded on EfficiencyOne's values of integrity, innovation, and partnership. The plan comprises four strategic priorities: achieve; innovate; champion; and grow.

The strategic plan also describes each of these priorities.

What it means to achieve:
- Put our customers first by providing exceptional value in all aspects of our work.

- Foster employee commitment by linking individual efforts to organizational values and priorities.

- Build organizational ability by developing skills that meet our future needs.

- Cultivate inspiring leaders by encouraging ongoing development and regular feedback.

What it means to innovate:

- Continuous improvement of businesses and governance practices, processes and relationships by embracing ideas, implementing them quickly and evaluating outcomes.

- Bring cost effective savings to our customers by continuing to adopt or define industry best practices.

- Provide our customers the best available and proven technologies by tapping into the world market of efficiency.

- Instill an efficiency first public mindset through education, research, partnerships and market transformation.

What it means to champion:
- Excel at the delivery of the Efficiency Nova Scotia franchise by achieving energy savings and meeting financial targets and goals.

- Contribute to Nova Scotia's environmental stewardship by promoting conservation and reducing energy waste and carbon pollution.

- Contribute to Nova Scotia's economic prosperity by growing Nova Scotia's efficiency industry and expanding its reach.

- Create a network of efficiency leaders in Nova Scotia through ongoing, meaningful communication.

- Put a spotlight on Nova Scotia's leadership in energy efficiency by sharing our story widely and effectively.

What it means to grow:
- Support and develop Nova Scotia's efficiency industry by attracting more businesses to the sector and helping them find opportunities for efficiency in new resources and in new jurisdictions.
- Bring our resource efficiency experience and expertise and all of its social, economic and environmental benefits to more people and more places."[16]

At the time of writing, the Province of Nova Scotia is in the process of developing a cap-and-trade program to contribute to global efforts on climate change. This cap-and-trade program will comply with the federal carbon pricing benchmark and is anticipated to regulate approximately 90% of GHG emitters.[17] This new system builds on Nova

Scotia's existing legal and regulatory frameworks to reduce emissions, some of which are listed below.

- 2007: The Environmental Goals and Sustainable Prosperity Act requires that, by 2020, Nova Scotia's GHG emissions be at least 10% below 1990 levels.[18]

- 2009: Nova Scotia's Climate Change Action Plan introduced a long-term target to reduce Nova Scotia's GHG emissions by up to 80% below 2009 levels.[19]

- 2009: Nova Scotia's Greenhouse Gas Emissions Regulations establish electricity GHG emission limits.[20]

- 2010: The Renewable Electricity Plan and Renewable Electricity Regulations require that 25% of Nova Scotia's electricity be supplied by renewable electricity sources by 2015, and 40% by 2020. [21]

ENS connects home and business owners with EE professionals through its ENS Trade Partner Network. To become a member of the Network, applicants must successfully complete two main training components, a general program overview workshop and customer service training. Training is carried out through a combination of online videos and testing.[22] The ENS General Participation Agreement compels partners to meet a number of requirements including conduct and ethical behavior and environmental responsibility standards, as well as maintaining workers' compensation insurance.

Partners who, as program subcontractors, represent ENS must meet stringent requirements. For example, installers for the Green Heat program component must hold Red Seal Certification with the Nova Scotia Apprenticeship Agency. In addition, the ENS website provides customers with a detailed online guide to selecting a contractor.[23]

ENS net incremental energy and peak demand savings undergo a rigorous, independent, multistage review every year, with slight differences between regulated franchise work (electrical energy and peak demand savings) and non-franchise work (non-electrical energy savings):

- Electrical Energy and Demand Savings:
  — EfficiencyOne staff calculate and track energy and demand savings for each program;

— Independent evaluators, whose services are retained by EfficiencyOne, visit customer sites, talk to participants as well as EfficiencyOne staff and contractors, establish their own savings estimates, and offer recommendations in a detailed final evaluation report which EfficiencyOne files with the UARB;

— An independent savings verification consultant, whose services are retained by the UARB, also conducts sample site visits, reviews the evaluation report, and prepares a verification report which is submitted to the UARB for review and approval; and

— These reports are publicly available on the UARB website.

• PNS Energy Savings:

— EfficiencyOne staff calculate and track energy savings for each project; and

— An independent evaluator, whose services are retained by EfficiencyOne, examines the findings and submits a final evaluation report which EfficiencyOne submits to the Province of Nova Scotia.

The following sections present the legal and regulatory framework and the building code.

LEGAL AND REGULATORY FRAMEWORK

The Nova Scotia Public Utilities Act requires NS Power to undertake reasonably available, cost-effective electricity efficiency and conservation activities in an effort to reduce costs for customers.[24]

The holder of the ENS franchise, currently EfficiencyOne, holds the exclusive right to supply NS Power with electricity efficiency and conservation services and is deemed a public utility when fulfilling franchise responsibilities.[25] EfficiencyOne is a federally incorporated non-profit corporation.

The franchise holder may also deliver other EE and conservation activities under the ENS brand with consent from the Province.

Since ENS franchise activities are regulated by the UARB, the latter approves DSM programs and expenditures and establishes performance requirements. Failure to comply with a performance requirement may be grounds for franchise termination.

Legislation requires EfficiencyOne and NS Power to enter into three-year agreements detailing the electricity efficiency and conservation activities EfficiencyOne provides and the amount NS Power pays for the supply of said activities.[26] The UARB has the authority to impose a final agreement when NS Power and EfficiencyOne are unable to come to terms.[27]

Despite the agreement, NS Power has no control over the design and implementation of DSM programs, thus ensuring the independence of the franchise holder.

Legal and Regulatory Timeline

2007: Nova Scotia's main electric utility, NS Power, prepares an Integrated Resource Plan stating that DSM programs are an integral, cost-effective strategy to meet emission reduction requirements and forecast increases in customer electricity loads. The UARB authorizes NS Power to develop and implement DSM programs.

2008: The Province of Nova Scotia commissions a public consultation and report to determine who should manage DSM programs. The consultation leads to the recommendation that an independent electricity efficiency entity be established that reports to a board of directors with oversight by the UARB.

2009: The Province of Nova Scotia passes the Efficiency Nova Scotia Corporation Act, thereby creating a non-profit independent administrator of EE and conservation services in Nova Scotia.

2010: Efficiency Nova Scotia Corporation begins operations in late 2010.

2014: The Province of Nova Scotia passes the Electricity Efficiency and Conservation Restructuring (2014) Act, thereby creating Canada's first electricity efficiency utility to be operated by EfficiencyOne which is a federally incorporated independent non-profit.

2015: ENS is a franchise operated by EfficiencyOne which is the official licensee of the Province of Nova Scotia.

Nova Scotia's Building Code Act provides the broad framework for the establishment of building code regulations in the province and delegates administration and enforcement duties to municipalities.[28]

EE is an important aspect of the Nova Scotia Building Code Regulations.[29] EE standards for medium to large non-residential or multi-unit residential buildings are imposed due to the adoption of the NECB, 2015.[30] This most recent version of the NECB was adopted by Nova Scotia effective April 1, 2017.[31]

EE regulations for houses and small buildings were established by incorporating the NBCC, 2015[32] in the aforementioned Nova Scotia Building Code Regulations. This most recent version of the NBCC was also adopted effective April 1, 2017.[33] New homes must meet a minimum EnerGuide rating of 80. The rating is provided by a certified auditor who models and verifies the measures installed during construction to ensure compliance, or through a prescriptive path that defines efficient measures which must be included in the design.

The Building Code Regulations also prescribe certain EE amendments and clarifications not stipulated in the NECB or NBCC, including the scope of buildings covered by the NBCC, the interpretation of appropriate climatic zones for Nova Scotia in accordance with the NECB, and considerations relating to the continuity of insulation for the NBCC.[34]

Nova Scotia requires certain consumer and commercial appliances and devices to meet specific EE or energy performance levels. The Energy-efficient Appliances Act[35] provides the legislative authority to prohibit the sale or lease of products that do not comply with the current Energy-efficient Appliances Regulations.[36]

Additionally, the Act and the Regulations were modified in 2011 and 2012 respectively to require all streetlights in Nova Scotia to be upgraded to LED technology. The current Regulations require that these conversions take place prior to 2020 for NS Power-owned lamps and before 2023 for all other lamps.[37]

Currently, the Energy-efficient Appliances Regulations [38] specify minimum energy performance criteria for 45 product categories. These criteria refer to specific standards mostly developed by the Canadian Standards Association.

As discussed previously, ENS net incremental energy and peak demand savings undergo rigorous multistage measurement, evaluation, and verification processes every year.

Although NS Power uses rates to manage system demand, such as its time-of-day rate for homes with electric thermal storage systems and interruptible rates for certain large industrial customers, demand response programs have not yet been implemented as part of DSM services provided by EfficiencyOne.

Under the Nova Scotia Municipal Government Act, municipal

councils may adopt bylaws enabling Property Assessed Clean Energy (PACE) financing programs.[39] Through ENS, EfficiencyOne offers complete PACE design, implementation and administration services on a not-for-profit basis.

Presently, five Nova Scotia municipalities manage PACE programs. PACE is a municipal financing mechanism that enables eligible homeowners to finance energy improvements through payments tied to a local improvement charge. Through this mechanism, all charges incurred for the home energy improvements are included in the PACE charges which attach to the property, not the individual.

If a homeowner decides to sell the home, the PACE charges carry over to the new homeowner. This ensures that both the payments and the energy savings remain with the people who enjoy the benefits of a comfortable and efficient home.

PACE helps homeowners overcome the major financial hurdle of investing in home energy improvements and allows for a repayment period of up to 10 years without penalty for paying the debt off early. All program service costs are paid by participants, and there are no costs to property owners who do not participate.

FINANCIAL MECHANISMS

For business, non-profit, and industrial customers, ENS provides a number of program incentives that include an on-bill financing option:

- Small Business Energy Solutions: Offers lighting replacements, and other energy efficient measures in small businesses, with more energy efficient units; provides audits and audit reports (if eligible), or a do-it-yourself option; covers up to 60% of the cost;

- Custom Retrofit and New Construction: The Custom Retrofit service incents building owners and operators to retrofit existing equipment with more energy efficient options, while the New Construction service incents developers to design and build new energy efficient facilities;

- Energy Management Information Systems: Incents industrial and large institutional organizations to design and implement energy management systems.

For residential customers, ENS offers low-interest financing (through a third-party lender) for Green Heat and Home Energy Assessment participants. ENS buys down a portion of the interest rate.

The electricity efficiency and conservation activities carried out by the ENS franchise are paid for by NS Power which recovers these funds from ratepayers. In accordance with the PNS contract, ENS also provides EE services to help low-income homeowners and renters save energy, which the Province of Nova Scotia funds.

In December 2015, EfficiencyOne incorporated EfficiencyOne Services under the Nova Scotia Companies Act, RSNS 1989, chapter 81. EfficiencyOne Services designs and delivers energy and resource efficiency services that are considered non-ENS franchise activities. As such, investments tied to EfficiencyOne Services transactions come from both public and private sources.

Chapter 9

Nunavut

ENERGY EFFICIENCY BACKGROUND

Nunavut is the newest and largest territory or province in Canada, as well as the most northern. In 1999, Nunavut separated from the Northwest Territories. According to the most recent statistics from the first quarter of 2017, Nunavut is the least populated province (37,462) [1] and has by far the lowest population density. Nunavut counts 25 communities with no interconnecting roads. The territory is also not connected by roads to the rest of Canada. Locals rely on air and marine transport for the delivery of goods and materials. Nunavut does not have an electricity grid that connects all communities, but the only energy utility, Qulliq Energy Corporation (QEC), operates standalone power systems sized to meet the demands of each municipality. The installed capacity produces energy 100% from fossil fuels.[2] Nunavut is completely reliant on imported fossil fuels for energy production.[3] Consequently, electricity prices are among the highest in North America.[4] EE and conservation measures are necessary to reduce fossil fuel dependence and GHG emissions. The emergence of sustainable communities that encourage EE and conservation measures can help mitigate the impacts of climate change, which are more tangible in Canada's northern regions.

The first concrete EE action in the then newly formed Nunavut came in 2003 when the Government of Nunavut (GN) released the Nunavut Climate Change Strategy. This document defined the early goals to cope with climate change impacts and reduce GHG emissions.

In 2005, QEC created the Nunavut Energy Center to run energy conservation programs. Designed to operate on federal funding, the Energy Center closed in 2009 after major federal cutbacks. Since then, energy conservation efforts have been taken solely by GN departments and agencies. Figure 9-1 below illustrates the timeline of EE and energy conservation initiatives.

In 2005, the Good Building Practices Guideline was released aiming to support constructors, architects, engineers and suppliers with a

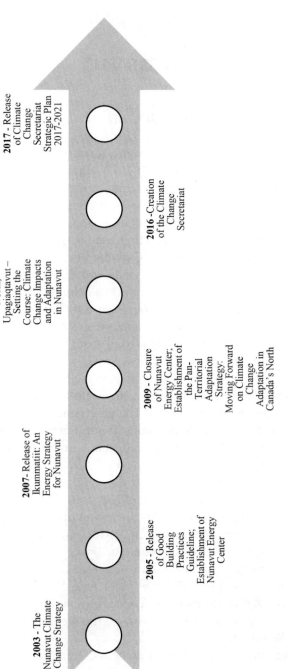

Figure 9-1: Timeline of Main Energy Efficiency Initiatives in Nunavut

comprehensive set of guidelines for constructing buildings in the North. However, the first real EE and conservation objectives were set by the Energy Strategy released in 2007, Ikummatiit. Alongside this plan, the City of Iqaluit established a pilot project for implementing EE measures in all GN buildings in the city by funding the Nunavut Energy Retrofit Program (NERP). This initiative also aimed to study the potential savings of similar measures in other Nunavut communities.

In 2009, the GN agreed to work alongside the governments of Yukon and the Northwest Territories on climate change by focusing on adaptation measures. The resulting Pan-Territorial Adaptation Strategy: Moving Forward on Climate Change Adaptation in Canada's North identifies the approaches for supporting climate change actions. [5] In 2009, the Northern Premiers' Forum, comprised of the territorial premiers of Nunavut, Northwest Territories and Yukon, committed to developing an inventory of available renewable energy resources. "This inventory describes the current state of renewable energy use in the territories, outlines actions being taken and describes policies under development to increase renewable energy use in the North. Finally, the inventory highlights the geographic and policy contexts faced by each territory that shape our distinct opportunities and challenges in the development of renewable energy."[6] The inventory was released in 2011 and is entitled Paths to a Renewable North.

In 2011, the GN also approved an updated strategy plan entitled Upagiaqtavut—Setting the Course: Climate Change Impacts and Adaptation in Nunavut. This plan establishes a framework for climate change impacts and adaptation initiatives. Upagiaqtavut enables the population to better adapt to current and future changes brought on by climate change. The objectives set in the strategy plan break down into four components:

- Partnership building;
- Research and monitoring impacts;
- Education and outreach;
- Government policy and planning.[7]

In 2016, the Energy Secretariat and the Department of Environment Climate Change Section merged to establish the new Climate Change Secretariat under the Department of Environment[8] with the objective of helping Nunavut adapt to the impacts of climate change.

In 2017, the Climate Change Secretariat released a five-year strategic plan (Climate Change Secretariat Strategic Plan 2017-2021). An annual work plan will be issued every year to reflect the then current situation. This Strategic Plan outlines four interconnected goals:

Goal 1—To ensure that the GN's climate change perspective and priorities are represented globally and its considerations are incorporated into National and International policies.

Goal 2—To demonstrate measureable progress towards climate change knowledge mobilization.

Goal 3—To demonstrate measureable progress towards climate change adaptation.

Goal 4—To achieve a measurable reduction in the rate of carbon emissions with minimal costs to Nunavummiut.[11]

STAKEHOLDERS

Government Departments and Agencies

The Climate Change Secretariat is responsible for the development, coordination and delivery of Nunavut's energy strategy. It develops and delivers energy-related programs and services. It aims at increasing the adoption of energy conservation practices, reducing energy waste and GHG emissions, as well as improving understanding of energy use. It also represents Nunavut in federal, provincial and territorial energy forums in which governments across Canada participate. Finally, the Secretariat is responsible for increasing awareness on energy use and climate change mitigation.[12]

The Department of Community and Government Services (CGS) of the GN collaborates with community governments to achieve better conditions for residents. CGS provides programs that support core municipal operations, as well as land and infrastructure development. CGS also supports government departments and agencies to efficiently manage their programs. It is one of the main government bodies contributing to the Ikummatiit Energy Strategy through different initiatives and information dissemination activities such as: Save 10—Employee Awareness Program; Energy Management Program of existing buildings owned by GN with a target of reducing energy consumption by 20%; amendment of the Best Practices Energy Guide; and finally the Government of Nunavut Leased Buildings program that serves to assess current EE barriers and adopt necessary changes.[13]

The Nunavut Housing Corporation (NHC), a public agency that provides approximately 70% of all housing in the territory, participates in EE efforts by providing incentives for EE upgrades through the delivery of the Public Housing program, GN Staff Housing program, and various homeowner support programs.

Except public housing (50% of the housing stock) and GN staff housing (13% of the housing stock), the NHC is responsible for the electricity and heating needs of 63% of all housing stock in Nunavut. In addition, NHC administers various homeowner support programs that encourage energy savings, such as the Home Renovation Program.[14] It also builds new houses using EE principles.

Energy Distributors

Qulliq Energy Corporation (QEC), the only generator, transmitter and distributor of electrical energy in Nunavut, is owned by the GN. QEC attends to the overall objectives provided by legislation and supports the Minister responsible for intergovernmental and regulatory issues. It is also mandated to manage capital projects and develop alternative energy generation sources.[15]

The Nunavut Energy Centre (NEC) was established in 2005 by the QEC to support municipalities, non-government organizations, businesses and individuals on energy management, efficiency enhancement, as well as conservation and renewable technology deployment. The Centre closed down in 2009 due to lack of federal funding.

POLICIES AND STRATEGIES

The first GN energy strategy was released in 2007, entitled Ikummatiit: An Energy Strategy for Nunavut, and provides guiding principles for energy policies, programs and activities until 2020. The policy actions set in this first energy strategy strive to attain the following objectives:

- Reduce reliance on imported fossil fuels;

- Reduce energy costs;

- Reduce energy-related emissions to decrease pollution and mitigate climate change impacts;

- Provide new business and employment opportunities.

The actions outlined in the energy strategy focus on four EE and conservation initiatives: Energy Awareness, Energy Education, Buildings and Equipment, and Transportation Energy.

Table 9-1: Energy Efficiency and Conservation Actions in Ikummatiit[16]

Energy Awareness	• Establishment of a resource centre to provide information and educational materials on energy efficiency and alternative energy sources; • A Point-of-Sale program to provide information on EE and alternative energy in stores; • The Save 10 program to promote EE awareness through newsletters, its website, and periodic information sessions and documents.
Energy Education	Provide training to: • Building operators and managers; • Building inspectors; • Energy efficiency and alternative energy trades training; • Pupils from kindergarten to Grade 12 through an energy curriculum.
Buildings and Equipment	• The new Energy Code for Retrofitting Existing Buildings; • Energy Management Program establishing a target of 20% reduction in energy consumption for government buildings; • Public Housing Energy Retrofit Program; • Reissuing of Good Building Practices Guideline including a chapter for Best Energy Practices for the northern territories; • Establishment of a working group to evaluate the barriers to EE and adopt necessary changes; • Residential, Commercial and Industrial Buildings Energy Retrofit Program to address the common barriers that private owners encounter when implementing EE measures; • Nunavut Energy Code for New Buildings; • Energy Star Labels for New Buildings; • New Commercial Building Program to ensure that new public buildings are designed to meet EE requirements; • Public Housing New Building Program to build new houses, 10-year housing program with the federal government, 725 new housing units to be built at an EE level that is 25% above the current code requirements; • New Building Support Program to help private owners access information and financial incentives; • Energy Efficiency Act to include further standards such as minimum EE standards for electric and fuel appliances, to restrict the sale of inefficient light bulbs; • Efficiency Incentives to discourage the sale of inefficient equipment; • Green Purchasing Policy; • Appliance Replacement Program.
Transportation Energy	• Transportation Energy Strategy to review transportation energy practices and develop a strategy for transportation energy use.

Three other strategies provide direction in Nunavut to cope with climate change:

- Upagiaqtavut—Setting the Course: Climate Change Impacts and Adaptation in Nunavut (2011), addressing the causes of climate change and adapting to its impacts.

- Pan-Territorial Adaptation Strategy: Moving Forward on Climate Change Adaptation (2009) in Canada's North outlines the impacts of climate change in the three Northern Territories and proposes strategies for collaborative actions while supporting specific initiatives to meet each territory's unique challenges.

- A collaborative document entitled Paths to a Renewable North, published in 2011,[17] outlines the common challenges, opportunities and issues affecting the Northern Territories, which include EE and conservation.

In October 2016, the federal government released its Pan-Canadian Framework on Clean Growth and Climate Change, under which all Canadian jurisdictions must have carbon pricing in effect by 2018. The GN and the Government of Canada have undertaken work to assess the implications of carbon pricing on energy costs and develop collaborative solutions. All revenues from carbon pricing will be retained by Nunavut. However, discussion is still underway on carbon taxing since Nunavut is not ready to introduce such taxation. The planned carbon pricing for 2018 is not applicable to Nunavut bacause it is highly dependent on diesel and renewable energy options are too expensive to implement quickly.[18]

LEGAL AND REGULATORY FRAMEWORK

In 2007, a bill entitled the Nunavut Energy Efficiency Act failed to pass in the Legislative Assembly. This act aimed to ban incandescent lights, but opposing members at the time argued that the federal government was already planning to ban these lights in 2012. No other EE legislation has been adopted in the territory.[19], [20]

FINANCIAL MECHANISMS

According to the website of Nunavut's energy system, the GN currently has no incentives or programs in support of renewable energy or EE technology. Support for renewable energy projects is, however, available through certain federal government programs.[21]

Only one program is available to Nunavut on the NRCan website—the Home Renovation Program.[22], [23] The program provides assistance to homeowners who repair or renovate their dwellings. Where applicants have already received assistance to renovate their home, they qualify for additional amounts for EE measures.

Chapter 10

Ontario

SECTION I—ELECTRICITY SECTOR

Background

In a world where the effects of climate change are becoming ever more apparent and widespread, using energy wisely has taken on greater importance in recent years. Conservation is the cleanest and most cost-effective energy resource available. In Ontario and elsewhere, it has become a key resource in helping customers save energy, control costs and reduce their environmental footprints.[i]

Over the past few decades, Ontario has used a number of approaches to drive greater demand-side participation and encourage different ways for energy consumers to reduce usage. To this day, the province's efforts in this area are broad, encompassing everything from traditional incentive-based conservation programs to training, education and support. Efforts also include direct load control, EE codes and standards, price-based models (e.g. time-of-use rates), demand response auctions, behind-the-meter (load-displacing) generation, as well as energy storage, just to name a few policies, programs and mechanisms.

Ontario Hydro and the Early Days of CDM

EE and conservation efforts in Ontario date back several decades, but the seed planted in the early 1980s by Ontario Hydro did not flourish until years later.

For the most part, the 1980s were dominated by supply-side initiatives,[1] including large-scale nuclear projects at the Pickering, Darlington and Bruce sites. Although some attention was paid to conservation and load shifting, and in 1982 targets of 1,000 megawatts (MW) for each

---

[i] For the purposes of this narrative, conservation and demand management (CDM) is used interchangeably with demand-side management (DSM). Both represent the full suite of customer-based energy solutions.

site were established by 2000, the province's supply situation meant that demand management received limited attention during that period.

When Ontario Hydro published its Demand/Supply Plan in 1989, it revealed a notable shift in the utility's long-term planning approach, such that resources on the supply side and the demand side were considered concurrently.[2] In addition, several new DSM programs were created. Although this plan was effectively shelved after the recession of the late 1980s and demand for electricity plummeted, especially from large industrial users, the Plan nonetheless established DSM as an important consideration in the context of energy adequacy and reliability in Ontario.

After the cancellation of Ontario Hydro DSM programs in 1993, the provincial Ministry of Energy offered a number of programs for commercial, industrial and residential consumers. However, these programs were discontinued in 1995 in preparation for deregulation. The creation of a wholesale electricity market was expected to incent consumers to use price signals as the basis for decisions on whether to consume electricity.[3]

Market Opening

The *Energy Competition Act*, 1998, opened the door to a competitive electricity market and effectively split Ontario Hydro into five successor companies, each with specific roles and responsibilities: (1) Ontario Power Generation assumed responsibility for power generation; (2) Hydro One took on transmission; (3) the Independent Electricity Market Operator, now known as the Independent Electricity System Operator (IESO), was in charge of market and system operations; (4) duties of the Ontario Electricity Financial Corporation related to debt and contract management; (5) and the Electrical Safety Authority handled training and safety standards.

In addition to enabling the creation of a wholesale electricity market, the Act triggered other structural changes. Whereas Ontario Hydro had been responsible for long-term planning, the restructuring process neither accounted for nor assigned that function to any organization since the market was not expected to require centralized planning.[4]

After several years of collaboration between government, stakeholders, Ontario Energy Board regulators and IESO technical experts, the Ontario wholesale electricity market opened on 1 May 2002, thereby introducing competition at the retail level.

Although local utilities engaged in some DSM measures in the early 2000s, coordination was limited. It was not until several years later, with the creation in 2005 of the Ontario Power Authority (OPA)—which merged with the IESO on 1 January 2015—that centralized power system planning and more coordinated energy conservation policies, programs and practices were established.

The creation of a wholesale market for electricity, coupled with extreme weather during the summer of 2002, resulted in volatile retail prices. Prices were eventually frozen in 2003 to reduce impacts on ratepayers. On the supply side, the market was not open long enough before the price freeze to attract new entrants willing to invest in new power generation resources. As a result of these and other factors, Ontario started experiencing energy shortfalls, and the IESO issued frequent public appeals to conserve energy.

Culture of Conservation

In June 2003, the government established the Electricity Conservation & Supply Task Force (ECSTF) whose mission was to develop an action plan to foster new power generation, promote conservation and enhance the reliability of the transmission grid.[5] Comprising 19 leaders from across the sector, the task force met on a weekly basis and heard expert testimony from dozens of stakeholders. Its final report, published in January 2004, was entitled *Tough Choices: Addressing Ontario's Power Needs.*

Among recommendations was a call to create a conservation culture in Ontario and an acknowledgment that conservation could play an important role in ensuring financial and environmental security, adequacy, affordability, reliability and sustainability. The report also explicitly noted that demand reduction should be given the opportunity to compete with supply-side alternatives and be evaluated on a level playing field.[6]

The ECSTF argued that to achieve the full potential that conservation offers, a lead organization needed to champion conservation efforts, coordinate the activities of the different market actors, and assess the relative costs and benefits of different programs to ensure consistency and alignment with government policy.[7] This call for a lead conservation agency was echoed by the Ontario Energy Board in a report to the Ministry of Energy issued in March 2004.[8]

The *Electricity Restructuring Act,* 2004, came into force in December of that year and enabled the creation of the OPA whose responsibilities included, among other things, forecasting electricity demand, planning for the power system, developing an integrated power system plan and facilitating its implementation, and finally promoting electricity conservation.

Conservation and Demand Management (CDM) and the IPSP

In addition to other functions, the OPA served as the focal point for conservation-related activities in the province. In its early days, it designed and launched a range of conservation programs delivered by local distribution companies (LDCs), as well as industry associations and private companies.[9]

As a government agency, the OPA is subject to directives from the Ministry of Energy, one of the most important of which, dated 13 June 2006, called for the development of an Integrated Power System Plan (IPSP) to meet a number of policy goals.[10]

This directive provided some degree of flexibility in how to achieve the first conservation goal of reducing 6,300 MW in peak demand by 2025. The directive broadly defined conservation to include EE codes and standards, as well as load-displacing generation such as small-scale natural gas-fired cogeneration and trigeneration facilities on the demand side of the meter.

In July 2006, the Ministry of Energy directed the creation of a three-year $400 million fund to help LDCs deliver CDM programs.[11] In addition to recognizing that LDCs play a legitimate role in delivering CDM, the Ministry also affirmed that stable multi-year funding should be provided to LDCs for program delivery and that the relationship between the OPA and LDCs should be managed contractually.[12]

A number of challenges arose associated with reintroducing electricity conservation programs after a decade of relative inactivity since the enactment of the *Energy Competition Act* in 1998. These included the need to rebuild the DSM program industry, disagreements on the proposed roles and responsibilities of various sectoral players, questions about the legitimacy of DSM as a resource, as well as alignment and cooperation between electricity and gas conservation initiatives to render conservation actions easier for customers.

Despite these initial hurdles, Ontario met its target of 1,350 MW in peak demand reduction by the end of 2007—the first step toward meeting the 2025 target.

The *Energy Conservation Leadership Act*, 2006, contained a number of provisions and requirements related to energy use by public agencies. The Act enabled the government to set specific targets for energy conservation for agencies and require the preparation of annual energy conservation plans. These plans, to be made publicly available, were expected to include itemized descriptions of energy-consuming technologies and operations, summaries of annual energy usage, descriptions of current and proposed activities and measures to conserve energy, as well as progress and achievement summaries in energy conservation.

Conservation and the Green Energy Act

The Legal and Regulatory Framework section in this chapter addresses the evolution of the Ontario legislative and regulatory framework for energy conservation in greater detail. Some of the most important changes stem from the enactment of the *Green Energy Act* (GEA), 2009, in May of that year. While much of this Act emphasized mechanisms to accelerate the adoption of renewable generation, it also included a renewed focus on conservation. Specifically, the GEA made energy conservation central to LDC operations by requiring that mandatory targets be met as a condition to maintaining their licenses.[13]

In addition to aggregating several existing conservation and EE policies, the GEA included important new provisions pertaining specifically to government-owned and/or government-managed facilities. It called on the provincial government to be guided by several conservation-oriented principles: clear and transparent energy consumption reporting; water use of and GHG emissions from government facilities; planning and designing government facilities to ensure the efficient use of energy and water; making environmentally and financially responsible investments in government facilities; and using renewable energy sources to provide energy for government facilities.[14] These requirements are further explained in the Energy Efficiency and the Public Sector section of this chapter.

The GEA also paved the way for the broader public sector, including municipalities, universities, schools and hospitals (also known as the MUSH sector) to track and report on energy consumption and GHG emissions, as prescribed in Ontario Regulation (O. Reg.) 397/11 which is discussed below in the Legal and Regulatory section of this chapter.

Save on Energy

The OPA collaborated with LDCs to deliver CDM programs under the 2011-2014 CDM Framework. In January 2011, a suite of programs was introduced under the saveONenergy brand. Administered by LDCs, these programs offered a variety of incentives for EE and demand management activities. They also served to engage homeowners and businesses on cost-effective ways to improve living and work spaces while saving on annual energy costs, help the province meet its energy needs, and reduce or defer the need to build new generation infrastructure.

Business programs were the primary source of energy savings among the 2011-2014 conservation programs, representing 58% of the overall savings achieved. During that period, the suite of business programs saved more than 4,000 GWh of energy and reduced peak demand by 389 MW, with the Retrofit program proportionally accounting for the greatest savings.[15]

Programs for residential customers, meanwhile, offered a range of incentives to retire old inefficient appliances and coupons to foster the purchase of energy efficient products such as compact fluorescent light bulbs, programmable thermostats, power bars, dimmer switches and motion sensors. Between 2011 and 2014, residential programs saved 1,169 GWh of energy and reduced peak demand by 253 MW—the equivalent of powering approximately 122,000 homes for one year.[16]

After 2014, Ontario transitioned to the 2015-2020 Conservation First Framework (CFF). This framework is described in the section below, along with the main conservation stakeholders as of 2017 and beyond.

Although the IESO plays an important role in administering formal CDM programs and catalyzing innovation through the Conservation Fund, LDC Innovation Fund and other mechanisms, other parties are carrying out a great deal of conservation work in Ontario, including consumers, technology companies, aggregators, incubators, academic institutions and other innovators.

For example, many Ontario LDCs are exploring other means of meeting customer energy needs, increasing resilience and security in their service area, and offering added value by introducing new products, services and technologies. The primary purpose of these efforts is not always to reduce total consumption or peak demand, yet some of the options being tested involve solar and storage systems, microgrids,

dynamic pricing models, enhanced telemetry and communication systems, sensors and controllers, electric vehicle integration, distribution
automation and others.

Tables 10-1 and 10-2 portray the active CDM programs in Ontario
(2015-2020), along with the roles of EE stakeholders.

Most energy conservation activities in Ontario stem from government policy. The Ministry of Energy sets broad energy policy, but
works closely with other ministries, including the Ministry of the Envi-

Table 10-1: CDM Programs in Ontario (2015-2020)

Province-wide Save on Energy Programs for Residential Customers	Province-wide Save on Energy Programs for Business Customers	Programs for Eligible Transmission-Connected Customers
Coupon (transitioning to Deal Days)	Retrofit	Industrial Accelerator Program (IAP) – includes all Save on Energy programs for business customers
Heating and Cooling	Small Business Lighting	Dispatchable Loads
New Home Construction	High-Performance New Construction	Demand Response Auction
Whole Home (delivered in coordination with gas utilities)	Existing Building Commissioning	Capacity-based Demand Response (transitional program for some existing loads)
	Process and Systems Upgrade	Industrial Conservation Incentive (ICI) – also available to eligible distribution-connected customers
	Energy Manager	
	Energy Performance Program (EPP)	
	Business Refrigeration Incentive (BRI)	
	Monitoring & Targeting	
	Training & Support	
	Audit	

Table 10-2: Organizations/Groups and Their Role in CDM

Organizations/Groups	Key CDM-oriented Responsibilities
Ministry of Energy	• Establishes the policy framework through the provincial long-term energy plan • Develops EE product standards • Sets reporting requirements • Demonstrates conservation leadership
Ministry of Municipal Affairs and Housing	• Sets building code requirements
Ontario Energy Board (OEB)	• Regulates the electricity and gas sectors • Approves licenses, rates and fees • Regulates gas DSM
Independent Electricity System Operator (IESO)	• Operates the electricity grid to balance supply and demand • Oversees provincial and regional electricity planning, which includes consideration of CDM to manage demand growth or defer the need for major new infrastructure in local areas • Administers the wholesale electricity markets • Administers the Demand Response (DR) Auction • Oversees the electricity CDM portfolio • Designs and delivers province-wide electricity CDM programs • Supports LDCs in designing electricity CDM programs • Provides funding for LDCs to pilot innovative CDM program ideas • Makes strategic CDM investments through the Conservation Fund • Provides central support services to LDCs • Works with gas utilities in areas of common interest • Manages CDM programs for transmission-connected customers • Manages centrally delivered CDM programs (Whole Home Program, Energy Performance Program and training incentives)
Local Distribution Companies (Electricity)	• Work with the IESO to design and deliver CDM electricity programs • Work directly with customers on CDM projects
Channel Partners	• Cross-sell CDM programs with other products, e.g. lighting and HVAC • Provide input and feedback that can be considered in the design of CDM programs • Provide energy services to customers • Deliver CDM programs under contract with LDCs and/or the IESO
Delivery Agents	• Deliver CDM programs to customers on behalf of LDCs and/or the IESO
Municipal Governments	• Develop corporate and community energy plans • Provide input for regional energy plans • Develop and implement policy • Demonstrate conservation leadership
Non-governmental Organizations (NGOs)	• Promote CDM programs and provide services to the community
Canadian Standards Association	• Develops EE product standards
Natural Resources Canada (NRCan)	• Administers federal Energy Efficiency Regulations which establish EE standards for energy-using products • Oversees the ENERGY STAR® initiative, a voluntary partnership between the Government of Canada and industry to promote EE
Local Distribution Companies (Gas)	• Design and deliver DSM programs for natural gas conservation • Work directly with customers on DSM projects
Indigenous Communities	• Develop community energy plans • Provide input for regional electricity plans

ronment and Climate Change, as well as other agencies, organizations and stakeholder groups.

As noted above, 2015 marked the start of a new approach to electricity CDM in Ontario with the launch of the 2015-2020 Conservation

First Framework (CFF). This framework had its origins in the government's 2013 Long-Term Energy Plan (LTEP) released on December 2, 2013. The 2013 LTEP went beyond the 2010 LTEP by setting a long-term conservation target of 30 terawatt hours (TWh) by 2032, which represented a 16% reduction in forecasted gross electricity demand.[17]

Conservation First Framework

While the 2015-2020 CFF excludes energy savings attributable to planned changes to building codes and product standards, it consolidates many of the province's planned actions in support of energy conservation goals over the six-year period. It emphasizes the importance of coordinated efforts at all stages of the energy planning process, as well as more effective collaboration among sector partners (especially the 60 plus LDCs operating in the province) and between gas and electricity sector stakeholders.

The overarching goal of the CFF is to reduce energy consumption by 7 TWh through conservation programs delivered by LDCs to residential and business customers across the province by December 31, 2020. An additional 1.7 TWh in energy savings are to be achieved through conservation projects undertaken by transmission-connected customers under the Industrial Accelerator Program (IAP) banner, for which the IESO is solely responsible.[18]

With a total budget of $2.2 billion, the CFF includes an expanded role for LDCs compared to previous frameworks. Another key difference with the 2011-2014 framework is that the CFF does not include demand reduction targets. Instead, each LDC is assigned a share of both the budget and the 7 TWh energy savings target, which they can pursue individually or in partnership with other LDCs.

To support LDCs in delivering conservation programs, the IESO provides central services including evaluation, measurement and verification (EM&V), market research, capacity building (including training), administrative infrastructure, and technical support. The IESO also designs and delivers certain province-wide CDM programs.

The IESO provides these tools, support and guidance to LDCs to help them meet their targets as outlined in the six-year CDM plans submitted to the IESO by each LDC. These plans enable LDCs to design and manage their own program offerings, providing greater flexibility to align conservation programs with local needs and more options for customers. The CFF also delivers long-term stable funding which provides LDCs greater funding certainty as they design and deliver the programs described in their CDM plans.

Through recommendations from customers, stakeholders, LDC working groups, and the EM&V process, program enhancements are continuously implemented to ensure current program offerings meet LDC, customer and system needs across the province.

Collaboration is a key focus of the CFF and helps to distinguish it from previous frameworks. With a view to maximizing efficiencies and minimizing costs, the IESO works closely with LDCs which are encouraged to partner with other LDCs to meet energy reduction targets and achieve efficiencies in program delivery.

Another key difference with previous frameworks is that only incremental energy savings are tracked under the CFF. Thus, LDCs are only credited for savings that persist to the end of the framework (2020). In practical terms, even if projects deliver savings for the first three years, they will not count toward the LDCs' targets if they do not persist to 2020. This approach of targeting allocation and attribution is presumably better for the ratepayer since LDCs exercise greater due diligence to determine which projects they fund.

With an emphasis on transparency, inclusivity, accountability, collaboration and efficiency, the Conservation First Implementation Committee (CFIC) supports the enhancement of existing province-wide programs and the development of such new programs to meet provincial conservation targets as cost-effectively as possible.

Comprising representatives from the IESO, LDCs, natural gas companies, the Ministry of Energy, the OEB and the Electricity Distributors Association (EDA), the CFIC establishes processes and provides oversight to LDC working groups that help in the design of cost-effective, customer-centric and results-driven programs.

In addition, the CFIC provides recommendations on the adoption of proposed new province-wide programs and potential program enhancements. It also monitors program results and LDC progress in achieving targets. The CFIC further helps develop new tools and services and provides limited guidance and input to the IESO on potential revisions to target and budget allocations among LDCs.[19]

One of the most important enhancements made in the development of the CFF was to increase LDC flexibility with respect to program design and delivery. Compared to the 2011-2014 framework, LDCs have substantially greater latitude to choose which programs work best in their own service territories.

As a result, in addition to offering standard province-wide programs, many LDCs have introduced local programming options that are closely aligned with their customer base, load profile and conservation priorities, including programs related to swimming pool efficiency, home energy reports, Wi-Fi-enabled smart thermostats, variable-frequency-drive pumps for agricultural uses and others.

Further information on CFF programs, including early results and key programs, is presented in the Policies and Strategies section further below in this chapter.

Market Transformation

To maximize the energy, employment, economic and environmental benefits of conservation, the IESO has adopted a market transformation approach to promoting, facilitating and driving conservation in Ontario. This process involves a strategic long-term view of the market and cost effectiveness of energy conservation measures. It includes interventions designed to identify and reduce barriers to participation, leverage new opportunities and new technologies, accelerate the adoption of energy efficient products, services, practices and measures, but also drive enduring changes in behavior—not just on the part of energy consumers but the market itself.

To accelerate this market transformation, many CDM programs delivered in Ontario involve not only LDCs, customers and the IESO, but also other channel allies. These third parties may include manufacturers, distributors, retailers, consultants, engineers, contractors, electricians, technologists and other individuals or organizations involved in the sale, service or installation of energy efficient equipment. As a result of the close relationship they share with customers, these trusted partners are becoming an increasingly important channel through which to more deeply engage electricity consumers, increase program participation and maximize energy savings.

IESO Conservation Fund

The IESO Conservation Fund is another important player in CDM, serving as a key incubator of new ideas, technologies and solutions. Created in 2005, the Conservation Fund supports innovative conservation technologies, practices, research and programs that offer the potential to deliver significant energy savings. It seeks projects that lead to large-scale transformation in the marketplace or a change in consumer behavior, or that demonstrate the energy saving potential of emerging technologies and research novel conservation concepts.

Since 2005, the Conservation Fund has provided financial support to more than 200 innovative energy projects across Ontario, some of which evolved from pilots to province-wide programs. The fund has invested more than $52 million, leveraging an additional $142 million from applicants and partners.

One of the more interesting projects financially supported by the Conservation Fund in recent years is the POWER.HOUSE program, which was initiated by a local utility (formally PowerStream, now known as Alectra) to study the feasibility of distributed energy resources as an alternative to the traditional wire solution for delivering electricity to customers. The funding has enabled this LDC to pilot a Virtual Power Plant using solar-storage technology installed at 20 residential customers on its service territory. The data generated will enable the LDC and the IESO to assess the customer, conservation, grid and utility business benefits of integrated solar-storage systems.[20]

Several other themes have dominated recent Conservation Fund applications and inquiries. For example, especially customers equipped with existing electric baseboard heating have expressed growing interest in heat pumps, while residential real estate developers perceive heat pumps as a key enabling technology for meeting net zero energy and other high-performance building standards in the Canadian climate.

Other recent proposals have focused on microgrids and district energy systems—an increasingly attractive model for municipalities, universities and other energy users with multiple buildings—a commitment to environmental sustainability and a critical need for an uninterrupted power supply. In addition to the potential cost savings associated with producing their own electricity, many of these consumers are keen to enhance their resilience against outages caused by extreme weather events and other circumstances.

Social benchmarking is another conservation tool with promise in Ontario. In the context of electricity usage, social benchmarking enables users to compare their electricity consumption with other similar users. In some cases, participating customers receive detailed home energy reports enabling them to monitor, track, compare and reduce energy use.

Three social benchmarking pilots that received support from the Conservation Fund in recent years enabled residential customers to compare their household energy consumption with that of similar homes in their neighborhood, which often resulted in a drop in consumption.

LDC Innovation Fund

The IESO also supports LDC-led program design and market testing of new initiatives through its central services budget. Testing and learning from small-scale pilot programs contribute to the success and cost effectiveness of launching new full-scale programs. Through the piloting process, LDCs market test the delivery mechanisms of and savings generated by new program offerings before including them in their CDM plans and budgets. This reduces program performance risks and allows the LDC to fine-tune delivery options on a smaller scale.

The IESO has funded more than 30 pilots through the LDC Innovation Fund and submissions are ongoing. Participating LDCs have developed pilots for residential, commercial, industrial, institutional and multi-residential customers using technologies such as smart thermostats, air source heat pumps, rooftops, ventilation and air conditioning system controllers, solar-powered attic fans, as well as electrically commutated motor (ECM) fan retrofits in existing home furnaces, just to name a few funded projects.

Industrial Conservation Initiative

As important as formal incented CDM programs and other funding mechanisms are, they do not represent the only options available to the IESO to reduce energy consumption. The IESO has a full suite of other tools available, including demand response (DR).

Amendments to the *Electricity Act*, 1998, enabled the creation of the Industrial Conservation Incentive (ICI) which is administered by the IESO. The ICI is an instance of voluntary DR. Customers that choose to participate in the program, referred to as Class A, are charged a global adjustment rate based on their percentage contribution to the top five peak demand hours in Ontario over a 12-month base period. The global adjustment rate is a financial mechanism used to recover the costs associated with conservation and generation investments; it is charged to all electricity consumers in Ontario in different ways. Class A customers that take action and shift their consumption away from peak hours significantly reduce their price exposure and total energy costs.

This initiative was introduced by the government in 2010 to encourage load shifting and provide opportunity for rate relief to large-volume industrial electricity customers. Initially, these customers were required to have an average monthly peak demand greater than 5 MW, but program eligibility was expanded several times with the threshold for participation now at 500 kilowatts (kW) for certain designated consumers.[ii]

In addition to offering advantages to customers, the ICI is making a material difference by reducing provincial peak demand and deferring the need to build costly new generation or transmission resources. Although other factors must be considered when making peak demand year-over-year comparisons—including the effects of embedded generation which reduces demand for grid-supplied energy and structural changes to the economy that have reduced the province's energy intensity—the ICI is clearly having the desired effect of incenting large load customers to shift consumption away from peak hours. In 2016, peak demand was reduced by approximately 1,000 MW as a result of actions taken by Class A participants through the ICI program.

Other Demand Response Programs

As market and system operator, the IESO considers DR an important tool that increases electricity supply reliability. By the early 2000s, the IESO had already introduced several DR programs. These included the: Emergency Demand Reduction Program; Hour Ahead Demand Response Program; Transitional Demand Response Program; and Emergency Load Reduction Program. The IESO also facilitated the integration of dispatchable loads into the Ontario energy market. The various programs comprised varying guiding principles, operational objectives, eligibility criteria and activation triggers, but all helped in different ways to support reliability when supply was strained.

Although the abovementioned programs were discontinued as needs changed, Ontario's dispatchable loads still reduce electricity demand by more than 450 MW.

DR has continued to evolve over the past decade and beyond. Through ministerial directives issued in 2005 and 2006, the OPA was required to procure up to 500 MW of DR, which resulted in the cre-

---

[ii] ICI eligibility requirements are established in O. Reg. 429/04 (https://www. ontario.ca/laws/regulation/040429).

ation of three DR programs: DR1; DR2; and DR3.[21] DR1, a voluntary load-shedding program, was discontinued, as was DR2, a mandatory load-shifting program.

DR3, a mandatory contract-based program, evolved into the Capacity-Based Demand Response (CBDR) program, a transitional program that brings DR3 participants into the wholesale electricity market. This program currently represents 159 MW of DR capacity and financially compensates both aggregated and individual load customers for making DR capacity available. The DR3 program also provides payments for reducing consumption at peak demand times when required. These resources are considered in the IESO economic dispatch process alongside bids from dispatchable load generators and offers from other resources.

The DR capacity of the CBDR program will decline over time as contracts expire and participation comes to an end. As this capacity expires, participants will have the opportunity to compete against other providers of DR capacity through the IESO DR auction, a competitive process through which demand-side resources are selected to reduce their electricity consumption during defined periods. The DR auction occurs annually and was established in December 2015.

Pursuant to a planned redesign of the Ontario wholesale market over the next four to six years, these demand-side resources are expected to compete against other resource types, including generation and imports, in a soon-to-be-established capacity auction that will meet Ontario's resource adequacy needs.

Development of an incremental capacity auction that can accommodate a range of resource types—on the supply and demand sides—is one of the priorities of the IESO Market Renewal program, a coordinated set of projects that will result in the fundamental redesign of Ontario's electricity markets. When fully implemented, it will provide increased opportunities for all stakeholders in Ontario's electricity market, with greater competition, flexibility and transparency resulting in efficiency gains for the entire sector, including customers and suppliers.

These efforts are part of the IESO long-term focus on delivering superior reliability performance in a changing environment and working toward a more efficient and sustainable marketplace.

The DR auction is already having the desired effect of attracting DR new entrants and lowering DR prices. Compared to the prices set during the first auction in 2015, clearing prices for the 2016 auction fell sharply, a 12% decrease for the summer commitment period and a 17% decrease for the winter commitment period, indicating meaningful competition between existing resources and new entrants.[22]

Conservation and Electricity Planning

Electricity planning in Ontario occurs at provincial, regional and local levels. While there are intersections between and among the various developed and implemented plans, each type of planning comprises its own requirements, considerations, processes and stakeholders.

At the provincial level, planning focuses on the flow of power across the province. This bulk electricity system delivers power to regional systems which in turn deliver power to transmission-connected customers and local distribution companies in the region. At this level, province-wide conservation programs and targets are considered in planning.

Regional planning is the process of identifying and meeting electricity needs for a specific local area and integrates local electricity priorities, as well as system and customer needs (e.g. community energy planning) with provincial policy directives, including those outlined in the LTEP.

Regional planning has been conducted on an as-needed basis for many years. In 2010, the OEB concluded that a structured approach to regional planning was required,[23] and this process was formalized in 2013.[24]

To ensure communities in Ontario have a safe reliable source of electricity supply, the IESO works closely with transmitters, distributors, customers, communities and other affected stakeholders to develop 20-year electricity plans for many regions across the province. These plans fulfill many needs, notably they: serve to identify electricity needs that could arise within a 20-year timeframe; consider infrastructure, generation, conservation programs, DR, distributed energy resources, and/or other provincial supply resources; and include near-term and longer-term recommended actions. Regional planning is an ongoing process, with each regional electricity plan revisited at least every five years.

As part of the regional planning process, LDCs, the IESO and electricity transmitters work collaboratively to explore potential CDM applications that could alleviate local constraints and defer (or eliminate) the need for costly investments in wire-based infrastructure.

There is no single demand-side measure that is effective in or appropriate to all areas. Every community's electrical infrastructure, load characteristics, building stock and customer demographics are different. As a result, the feasibility of CDM solutions must be evaluated on a site and case-specific basis. What works in one area may not be viable or even appropriate elsewhere.

Currently, limited data is available to assess the cost and viability of targeted demand-side solutions such as demand response, energy storage, small-scale generation and other distributed energy resources at the local level (i.e. at the transformer station). With the support of the Conservation Fund, local potential studies which serve to assess the achievable conservation potential of a particular area are being carried out in several parts of the province.

When completed, these local potential studies should provide valuable data that will be leveraged by LDCs and the IESO to determine whether CDM solutions are technically feasible and cost-competitive in those specific areas.

Local planning, by contrast, is typically carried out by the LDC and focuses on delivering power from the regional system to individual homes and businesses. As part of this process, LDCs work to achieve their respective conservation targets and assist municipalities in community energy planning.

Leveraging the Value of Smart Meter Data

Access to good data is a growing focus in Ontario. Similarly to other jurisdictions around the world investing in smart grid infrastructure, the increased digitization of the Ontario grid is beginning to enable better coordination, communication, analytical capability and asset management.

Ontario took an early lead in the development of a smart grid by mandating the installation of smart meters and the transition to time-of-use prices. Approximately 4.7 million smart meters were installed in homes and small businesses earlier this decade, which are now sending enormous quantities of energy consumption data to a centralized database known as the Meter Data Management/Repository (MDM/R) which is managed by the IESO in fulfillment of its role as the Smart Metering Entity.[iii]

[iii] The IESO's role as SME is established in O. Reg. 393/07 (https://www.ontario.ca/laws/regulation/070393).

Although most LDCs in the province use the MDM/R for billing purposes, the true value of the data contained therein remains largely untapped. Upon a directive from the OEB,[25] the IESO has begun collecting enhanced data and is working with stakeholders to explore ways to extract the full value of Ontario's smart meter data while ensuring that customer privacy is protected.

The Data Strategy Advisory Council was created in mid-2017 to provide input on IESO's development of products and processes aimed at providing third-party access to meter data. Comprised of representatives from LDCs, municipalities, gas and water companies, as well as other service providers, this group will help define the path forward. In addition to identifying opportunities for the public and private sectors to develop new products, services and technologies, the data captured in the MDM/R could be used to further refine CDM program offerings.

Codes and Standards

Turning to Ontario CDM activity that is not incented, administered or overseen by the IESO, codes and standards are other important contributors to energy conservation, representing approximately one third of total energy savings. Appliance and product standards[iv] apply to household and commercial appliances such as water heaters, furnaces and other space heating equipment, air conditioners, lamps and other lighting products, motors, pumps, electronic equipment and other related products. Importantly, the federal government also regulates equipment standards.

For new houses and large buildings, the EE requirements contained in the Ontario *Building Code*, administered by the Ministry of Municipal Affairs and Housing, are among the most rigorous in North America. With every iteration of the Code, energy performance requirements have been strengthened for new constructions. Two CDM programs, New Home Construction and High Performance New Construction, offered through the CFF are designed around the Code and provide incentives to customers that upgrade their homes/buildings beyond Code requirements.

Unlike new buildings, there are no EE requirements for retrofitting

---

[iv] Appliance and product standards in Ontario are regulated by O. Reg. 404/12 under the Green Energy Act, 2009 (https://www.ontario.ca/laws/regulation/120404).

existing buildings. Like other provinces and territories, Ontario is monitoring developments at the federal level with interest. EE requirements are being considered for existing buildings and, once available, a National Energy Retrofit Code for Buildings may be voluntarily adopted by provinces as part of their EE regulations.

Although the EE requirements contained in the Ontario *Building Code* are progressive by North American standards, they fall short of standards adopted by leading European jurisdictions. As indicated in the Government of Ontario 2016 *Climate Change Action Plan* however, the Ontario *Building Code* is moving toward net-zero carbon emission homes and small buildings by 2030. [26] Further information on codes and standards is outlined below in the Legal and Regulatory Framework section of this chapter.

Policies and Strategies

Ontario conservation goals, targets and priorities are factored into energy policy and planning at different stages—most importantly during the long-term planning process. The Ministry of Energy is responsible for issuing a long-term energy plan[v] that lays out the Government of Ontario's goals and objectives with respect to energy—a process that replaced the Integrated Power System Plan framework under which the IESO, OEB and other stakeholders previously operated.

Ontario produced its first LTEP in 2010, which was reviewed and updated in 2013 with input from residents, businesses and other stakeholders in communities across Ontario. The government also issued a discussion paper in 2013 entitled *Conservation First: A Renewed Vision for Energy Conservation in Ontario*. Therein, the government reaffirms its commitment to prioritizing conservation in energy planning, stating that conservation should be the first resource considered to meet Ontario's electricity needs.[27]

The latest LTEP was released in the fall of 2017. The government has indicated that five principles will guide decision making related to the 2017 plan: affordability; reliability; clean energy; community and Indigenous engagement; and CDM.

---

[v] The Ministry of Energy's role is enshrined in the Energy Statute Law Amendment, S. O. 2016 c. 10 (also known as Bill 135 [https://www.ontario.ca/laws/statute/S16010]).

Ontario Planning Outlook

As required by legislation, the IESO provided a technical report to inform government planning and prioritization. Published on 1 September 2016, the IESO *Ontario Planning Outlook* (OPO)[28] is a technical report that provides a 10-year (2005-2015) review of and a 20-year outlook (2016-2035) for Ontario's electricity system. This sweeping report covers a broad range of considerations including costs, CDM, supply resources (including energy storage), capacity, reliability, market and system operations, transmission and distribution, and finally electricity sector GHG emissions.

The OPO also examines how conservation and other low-carbon resources can support reliable and efficient operations over a range of possible demand outlooks. Each of four demand scenarios assumes the achievement of three key targets set in the 2013 LTEP:

• 30 TWh in total energy savings by 2032;

• 7 TWh in energy savings through LDC programs delivered under the CFF and 1.7 TWh in savings through the IAP by 2020; and

• Demand response to meet 10% of peak demand by 2025.

The OPO delves into the many factors expected to impact electricity demand over the next 20 years, including population growth, commercial and industrial sector activity, electric vehicle penetration levels, and the electrification of space heating, water heating and transportation, among others.

Furthermore, the OPO predicts that by 2035 almost half of all energy savings in Ontario will stem from updated building codes and equipment standards which deliver EE with a high degree of certainty and require no ratepayer incentives. Codes and standards implemented to date will continue generating savings, increasing to approximately 7.9 TWh by 2035.

Planned and anticipated codes and standards should save an additional 7 TWh over the same period, with the vast majority of energy savings generated by the commercial and residential sectors. In total, existing and anticipated building codes and equipment standards are projected to yield approximately 14.9 TWh in savings by 2035, with almost 9.6 TWh generated from equipment efficiency standards and about 5.3 TWh from building code changes.[29]

The Ontario Decarbonized Supply Mix

Ontario has already made great strides toward reducing the impacts of human activity on the environment, including the elimination of all coal-fired power generation by 2014—the single largest climate change initiative undertaken in North America and an important contributor to the province achieving its ambitious 2014 emission reduction target of 6% below 1990 levels.[30]

Building on a solid foundation that includes nuclear generation and renewable resources such as hydroelectric, wind and solar units, the Ontario electricity supply mix is already largely decarbonized. Since the supply shortfalls of the early 2000s, more than $35 billion has been invested in cleaner generation, coupled with $16 billion in transmission and distribution upgrades.[31]

Electricity generation in Ontario produces little GHG emissions. Transmission-connected natural gas-fired generators provided only 9% of Ontario's electricity in 2016.[32] While the contribution of gas-based energy generation is relatively minor in absolute MW, it will continue to play an important role in Ontario for the foreseeable future, providing flexibility and enabling the IESO to quickly respond to unexpected changes in supply and/or demand. In addition to providing energy and capacity, this flexibility enables the use of gas generators to deliver a variety of ancillary services essential to operating a well-balanced, reliable and cost-effective power system.

Going forward, the government has signaled its intention to continue aligning energy and climate change policies with one another. The *Climate Change Mitigation and Low-Carbon Economy Act*, 2016 (Bill 172) was adopted in May of that year. The legislation sets the framework for climate action and establishes the following statutory GHG emission targets.[33]

- A reduction of 15% below 1990 levels by the end of 2020

- A reduction of 37% below 1990 levels by the end of 2030

- A reduction of 80% below 1990 levels by the end of 2050

Approximately one month later, the government released its *Climate Change Action Plan* which identified more than 90 actions to be implemented over the ensuing five years in an effort to fight climate change, reduce GHG emissions and transition to a low-carbon economy. Many of these actions will impact how residential, commercial, industrial and institutional customers perceive and use energy.[34]

Collaboration with Gas Utilities

The Climate Change Action Plan includes a $325-million commitment to the Green Investment Fund to finance local environmental organizations, as well as the installation of more electric vehicle charging stations and retrofits to social housing developments. The funds also help homeowners use less energy, businesses reduce emissions, and provide Indigenous communities with training, tools and infrastructure to address climate change.[35]

Of that amount, $100 million is earmarked to help homeowners reduce their energy bills and cut GHG emissions through electricity-focused enhancements by expanding the Enbridge Gas Home Energy Conservation program and Union Gas Home Reno Rebate program. These multi-fuel province-wide programs constitute a coordinated effort by the IESO and the two largest gas distributors, Enbridge and Union Gas. The programs provide residential consumers with a single point of entry and a streamlined approach to identifying and financing home EE improvements. Enhancements have expanded program eligibility to electrically heated homes and provide an additional in-home assessment of electricity use as well as offer new incentives to reduce electricity consumption, including ENERGY STAR® qualified air conditioners and appliances, gas furnaces with electronic control modules and air source heat pumps.

Audits provide homeowners with a holistic view of their home, including energy requirements and savings opportunities. Armed with this information, homeowners can then prioritize their spending on the EE measures with the most savings potential. The program also provides larger incentives to drive investments in high-cost and high-value equipment such as air source heat pumps which have steep upfront costs but offer substantial energy savings potential. The electricity-related enhancements to these programs were launched in May 2017 and 22,000 households were expected to participate in these program components by the end of that year.

In addition to reducing the electricity consumption of electrically heated homes by approximately 5,000 kWh per year by installing energy efficient upgrades, participating households should also benefit from improved home comfort and increased resale value.

Addressing Climate Change

The *Climate Change Mitigation and Low-Carbon Economy Act*, 2016, also enabled the creation of a cap-and-trade program. The proceeds from cap and trade will be used to finance initiatives intended to reduce GHG emissions and will be administered by a new agency called the Ontario Climate Change Solutions Deployment Corporation. Established in February 2017, this agency is more commonly known as the Green Ontario Fund.[36]

This green fund was created to stimulate the development of commercially available technologies that reduce GHG emissions from buildings or the production of goods. Its focus is on providing information, services and incentives to individuals, stimulating private-sector financing, as well as identifying and reducing market barriers that currently inhibit the deployment of low-carbon technologies.

Over the spring and summer of 2017, the IESO began working closely with the Ministry of Energy and the Ministry of the Environment and Climate Change to support and prepare for the launch of the new agency.[37] This work is ongoing. An initial portfolio of programs was launched in fall 2017 by first targeting residential consumers and then business customers. Leveraging IESO expertise in designing and delivering appealing customer-focused CDM programs, these early initiatives launched by the Green Ontario Fund will build on existing infrastructure and offer a one-window approach to customer service and simple processes that should foster high participation rates.

As Canada moves toward a lower carbon economy, Ontario is starting from a place of relative strength by already having a largely decarbonized supply mix. Renewable energy now comprises 40% of Ontario's installed capacity and generates approximately one third of the electricity produced in the province. Combined with nuclear resources, which account for one third of Ontario's installed capacity and produce nearly 60% of its electricity, these non-fossil fuel sources now generate approximately 90% of Ontario's electricity.[38]

Electrification—whether for space/water heating in buildings, railways and public transit, or passenger and commercial vehicles—is expected to provide an important opportunity for Ontario to effect a sensible, strategic, and efficient transition.

Conservation First Framework—Key Programs and Results

As noted further above, the CFF has transformed the incented

CDM landscape in Ontario. LDCs are making good progress toward their 2020 targets, and the LDC CDM portfolio remains cost effective. Similarly, many of Ontario's 59 transmission-connected customers, among the province's largest employers, have undertaken multi-year IAP-funded projects to drive operational efficiencies. Like the LDC-delivered program portfolio, the IAP remains cost effective, although it may be several years before these large complex projects fully deliver expected conservation results. Both sets of programs are currently delivered at less than $0.04/kWh, thus establishing conservation as the cheapest form of energy supply.

LDC-delivered programs are managed and promoted under the Save on Energy (SOE) brand and include options for residential and business customers (commercial, institutional and low-volume industrial users). Programs for business customers deliver the lion's share of results. In 2015-2016, these business programs delivered 70% of total energy savings against the 2020 target. Residential programs, by contrast, represent 29% of total savings, though this percentage is on the rise in large part due to the success of the Coupon program targeting residential EE products.

In-store rebates are popular with residential consumers, with 14.7 million coupons redeemed across the province, representing 24.8 million energy efficient products purchased in 2015-2016.[39] Results in 2016 indicated a notable improvement over 2015, with a 92% increase in coupon redemptions and a 119% increase in energy efficient products sold.

Some of this growth is to be attributed to recent price drops of energy efficient products compared to inefficient alternatives, which rendered upgrades more financially palatable than in previous years. That said, the success of the Coupon program is also evidence that Ontarians understand the full spectrum of benefits that EE delivers.

To build on these results, the program model was modified in late 2017. Instead of relying on coupons, the new Deal Days point-of-sale discount program is expected to further increase engagement with consumers by enabling them to benefit from incentives when making online energy efficient purchases while simultaneously reducing administrative costs and challenges for LDCs and the IESO.

For business customers, the most important program in Ontario is the Retrofit program. Delivering approximately 52% of total energy savings tracked thus far under the 2015-2020 Conservation First Framework and the 2011-2014 framework, the Retrofit program provides incentives for participants to replace outdated inefficient equipment with high-efficiency alternatives and install new control systems that improve operational processes.

While the program is subject to continuous improvement and change, eligible measures currently include lighting retrofits and controls, HVAC redesign, chiller replacements, variable speed drives, building envelope improvements, as well as new operating systems, procedures and/or equipment including energy management systems, sensors and controls, metering equipment and related communication systems.

The Retrofit program truly delivers for Ontario businesses. It has broad applicability and can be implemented by most types of commercial, industrial and institutional consumers. It also includes both prescriptive and custom tracks for maximum flexibility. When the next iteration of the program is brought to market, it is expected to reflect the natural evolution of products, technologies, applications and prices.

Training and Support

For optimal results, funding to purchase and install energy efficient equipment should be complemented by training that enables users to operate the equipment more effectively and identify other potential improvements in facilities. In Ontario, building market capacity through training and support is becoming as much a priority as incenting the purchase and installation of energy efficient products.

As a market transformation initiative, the SOE Training and Support incentives are designed to enhance CDM-oriented knowledge and skills among LDCs, channel partners, customer employees and other individuals involved in implementing SOE-funded projects. Incented training courses enable participants to identify operational savings opportunities, increase employee engagement in energy matters, integrate new technologies and best practices, and secure executive approval to implement proposed changes. As the capacity of Ontario businesses and institutions grows and since EE is considered a standard practice for planning and budgeting purposes, the need for ratepayer-funded incentive programs is expected to decrease.

Energy Manager Programs

Since 2011, Ontario and the IESO have also provided direct financial support to large electricity consumers to hire energy managers to increase participation in existing programs and foster the implementation of non-incented EE projects as well as the adoption of operational and behavioral measures. Energy managers are active in all sectors of the Ontario economy, notably in industrial sites, office buildings, retail complexes, multi-unit residential properties, hospitals, universities and municipal buildings.

Inspired by BC Hydro's pioneering Energy Manager Program and informed by other pilot projects through the IESO Conservation Fund, the IESO Energy Manager Initiative provides incentives to hire energy managers who work full-time at participating sites.

Over the 2015-2016 period, the Energy Manager Initiative produced more than 38 million kWh in net verified energy savings and more than 600 kW in net verified peak demand reduction, both persisting to 2020.

Interest from large-volume transmission-connected customers eventually prompted the expansion of the IAP to provide energy manager funding to eligible customers such as steel mills, petrochemical facilities, automotive factories, and mines, among others. As of March 2017, more than 17 IAP customers—representing almost one third of all eligible customers—participated in the IESO IAP Energy Manager Initiative.

In the spirit of providing conservation opportunities that meet customer needs and expectations, the IESO also worked closely with LDCs, customers and service providers to create an Energy Manager component specifically for customers with facilities located on multiple LDC service territories. For multi-distributor customers, this program offers administrative simplicity through a single contract and a single point of contact with the IESO for activities across the province.

To maximize their chances of success in identifying and implementing EE improvements, IESO-funded energy managers have access to a broad range of tools and support including a private online discussion forum, monthly newsletters, in-person events, training courses on presentation skills and financial basics, as well as research among other resources.

Energy Performance Program

Multi-site customers were the beneficiaries of another program designed and implemented in 2016: the Energy Performance Program (EPP). After a successful Conservation Fund pilot project undertaken by a large grocery chain, the program was formally launched in December

2016. Designed to be highly flexible, the program enables participants to choose which conservation measures are most appropriate and offer the greatest energy savings potential to their businesses.

The EPP provides commercial and institutional customers with facilities located in multiple service territories the opportunity to receive EE incentives on a whole-building pay-for-performance basis, thereby leveraging building interval meter data—the first such program of this scale and scope in North America.

The unique EPP structure promotes operational and behavioral changes to achieve EE, along with capital investment projects, and reduces the administrative burden on large multi-site customers by enabling them to apply for larger incentives through one window for many different facilities. These customers must sustain achieved electricity savings for multiple years to receive the larger incentives. Although the program is centrally administered by the IESO, savings are credited to LDCs and count toward their CDM plan targets.

Evaluation, Measurement and Verification

Rigorous and independent EM&V of conservation results are essential to credibility. Evidence-based analysis helps establish CDM as a valuable resource for meeting Ontario's energy needs by providing answers to key questions. Can conservation reliably deliver energy savings? Is conservation more cost effective than supply-side alternatives? Do the benefits of conservation outweigh the costs? Finally, is conservation a cost-effective and appropriate use of public funds?

EM&V provides essential information to policymakers, system planners and program administrators who use it to develop long-term plans, optimize program performance and determine whether energy savings and demand reduction targets are being met. This information is also used to develop a reliable net savings estimate defined as the energy savings attributable to or resulting from program sponsored efforts as distinguished from other measures.[40]

Cost effectiveness is a priority in Ontario and has broad implications for program planning, design and implementation to ensure ratepayer dollars are invested wisely. In general terms, conservation is considered cost effective when the lifecycle costs of implementing conservation programs and actions are lower than the benefits (e.g. avoided costs of new generation, transmission or distribution).

Ontario has drawn from existing EM&V best practices developed in other jurisdictions and adapted them to ensure they are meaningful and relevant. In some cases, the province has used established benchmarks as a foundation for more rigorous EM&V protocols and processes. As a result of data-collection activities thereof, stakeholders have access to new and strategic information that serves to ensure program and portfolio effectiveness.

To continue meeting established policy objectives and other criteria, the IESO evaluates all conservation programs every year, compared to many other jurisdictions that only evaluate programs every few years and/or integrate spot checks into their EM&V processes. Regularly evaluating programs deepens the collective understanding of what is driving a market, enables administrators to identify potential enhancements and reveals which programs deliver the greatest benefit to customers and the system as a whole. All evaluation results are made public in support of commitment to transparency.

Cost-effectiveness metrics are used to assess CDM during planning stages and as part of EM&V processes. CDM is assessed at various levels: measure; program; or portfolio. The measure is the most granular level of CDM and is comprised in the conservation technologies, products or actions implemented by participants. The program is defined as a collection of measures and/or activities targeting, for example, a particular end use (e.g. lighting) or customer type (e.g. small commercial). The portfolio is a collection of programs.[41]

The IESO uses numerous tests when evaluating the cost effectiveness of its CDM activities. In addition to assessing individual programs, the IESO evaluates them on a broad portfolio-wide basis that factors in other policy drivers and social objectives. The most meaningful test is the Total Resource Cost (TRC) test, which compares the costs incurred to design and deliver conservation programs, as well as customer costs, along with the avoided supply-side resource costs including generation, transmission or distribution options. The strength of the TRC test is that it provides a holistic perspective that examines the costs and benefits to both local utilities and participating customers.[42]

Other common tests used to assess conservation programs include the Program Administrator Cost (PAC) which is focused solely on the costs to program administrators (e.g. LDCs or the IESO) and the Levelized Unit Electricity Cost (LUEC) which normalizes the costs incurred to design and deliver CDM programs on a unit basis to enable comparison with supply-side alternatives.[43]

Legal and Regulatory Framework

Codes and Standards

The *Ontario Energy Efficiency Act*, 1990, granted the province the authority to require MEPS on the sale of certain energy-consuming products. Two years later, the federal Energy Efficiency Act enabled the federal government to require MEPS on products traded across provincial and international borders. Both pieces of legislation, and the associated codes and standards, have fostered substantial energy savings in Ontario in the intervening years.

Regulating building construction is a provincial jurisdiction. Ontario's first *Building Code Act* was proclaimed in 1974 with the first *Building Code* regulation taking effect the following year.[44] It is updated roughly every five years, with the most recent *Building Code* (2012) coming into force on 1 January 2014. The code is administered by the Building and Development Branch of the Ontario Ministry of Municipal Affairs and Housing. While the code is focused on ensuring public safety in newly constructed buildings, it also supports the government's commitment to energy conservation, barrier-free accessibility and economic development.

To the extent possible, the province attempts to harmonize Ontario's *Building Code* with model national construction codes such as the NBCC and the NECB which were designed to complement provincial/territorial building codes. These two voluntary Canadian codes set out minimum EE requirements that may be incorporated, in whole or in part, into provincial and territorial legislation and codes or, alternatively, used as guidelines for constructing new energy efficient buildings.[45]

As noted above, the *Green Energy Act*, 2009 (GEA)—also known as the *Green Energy and Green Economy Act*—is better recognized for encouraging renewable generation than promoting energy conservation. However, it also introduced EE standards for a number of household and commercial appliances through O. Reg. 404/12.

Energy Efficiency and the Public Sector

Amendments to the GEA introduced in 2012 brought other changes. Most notably, buildings that fall under broader public sector (BPS) ownership—municipal facilities, universities/colleges, schools, hospitals and the like—became subject to new energy consumption and GHG reporting requirements.

As established under O. Reg. 397/11, affected groups in the BPS are required to publicly report energy use and GHG emissions and establish five-year CDM plans. The Environmental Commissioner of Ontario (ECO) 2015-2016 Annual Energy Conservation Progress Report suggests the BPS-owned and/or leased portfolio represents about 590 million square feet of floor space across 15,000 facilities with total energy consumption of 18.6 billion equivalent kWh (ekWh)/year.[46]

While voluntary programs can and do deliver useful results, mandatory energy reporting programs are generally considered more effective in driving long-term persistent energy performance improvements. Mandatory reporting regimes entail heightened scrutiny of energy consumption patterns not only on the part of building owners who are required to report relevant data, but also residents, special interest groups, government bodies, and even independent third parties seeking to improve on existing results through new approaches or technologies.

With several years of consumption data now available, the 2015-2016 ECO report declares that energy use in Ontario public buildings still offers "significant potential" for EE improvements. When fully implemented, these improvements will produce many benefits, including reduced operating costs, less demand on energy infrastructure, more green jobs and reduced GHG emissions.[47]

As important as energy consumption is in the BPS, those properties represent a fraction of the untapped conservation potential across the province. Provincial public buildings, whether owned or leased by the government itself or by the BPS, account for just 8% of the energy used in all Ontario buildings.[48]

O. Reg. 20/17 was passed under the *Green Energy Act*, 2009 in February 2017. It introduced new energy consumption and water use reporting requirements for certain privately-owned properties exceeding 50,000 square feet.[49] These include office buildings and other commercial and multi-unit residential properties with more than 10 units, as well as retail stores and some industrial properties.[50]

For commercial property owners and managers, benchmarking energy consumption enables organizations to compare facilities within their own portfolio as well as to facilities across the country. Benchmarks help identify which properties present the greatest savings opportunity. Moreover, depending on how facilities measure up to their peers, results can greatly motivate property owners to intensify their EE efforts.

In this context, the IESO has been a strong supporter of the EN-ERGY STAR® Portfolio Manager benchmarking tool offered by NRCan. Launched in 2013, this industry-leading interactive energy management tool enables commercial and institutional building owners and managers to upload their building and utility information to an online platform that tracks energy performance over time and enables meaningful comparisons to similar buildings.

This benchmarking tool provides weather-normalized energy-use intensity values, GHG emission metrics and reporting features that help users track trends over time. Furthermore, a 1-100 energy performance scale permits eligible building types to be scored for certification. For example, a score of 75 indicates that a facility is performing better than 75% of its peers and is eligible for certification. Importantly, the Ontario Energy and Water Reporting and Benchmarking (EWRB) regulation requires the use of Portfolio Manager for reporting purposes.

Under the aforementioned O. Reg. 20/17, reporting is being phased in over three years (2018-2020) according to property size from largest to smallest. Although some large commercial property owners have been using Portfolio Manager for several years and found ways to leverage what they have learned from benchmarking against their peers, this process is new for many participants including large retail facility, warehouse and condominium owners.

The Ministry of Energy has established a stakeholder working group to support the implementation of its Large Building EWRB initiative. Through IESO-supported initiatives such as the race2reduce and Housing Services Corporations Utility Management programs, a significant number of private and public property owners have already voluntarily begun benchmarking buildings using Portfolio Manager, thereby helping to reinforce benchmarking as a best practice and establish Portfolio Manager as the default tool in Ontario.

In support of the 2007 Go Green Action Plan (GGAP), the Ontario Ministry of Infrastructure (MOI) committed to reducing GHG emissions through energy conservation measures. The GGAP outlined overall Ontario Public Service (OPS) targets in a number of areas including OPS building emissions.

Ontario has met GHG emission reduction targets through extensive energy retrofits in government-owned buildings and energy conservation measures. To help achieve the overall GGAP targets, the MOI committed to reducing building emissions by 27% compared to 2006 levels by 2020. Importantly, the MOI portfolio had already achieved a 19% reduction target by 2014.

Annual energy and GHG reporting requirements are outlined in the Ontario Facilities Energy Consumption Directive (as per the *Green Energy Act*). Annual energy and GHG reporting requirements are divided as follows:

• The 5-Year Energy Plan reporting on target buildings and initiatives for MOI buildings;

• The Enterprise-Wide GHG Summary which includes the MOI-managed portfolio as well as custodial ministries that have buildings managed by the ministry directly and fall outside the purview of the MOI.

In support of the building-related GHG emission targets for 2014 and 2020 (19 and 27% respectively), an electricity reduction target was set at 20% compared to 2002 levels by 2012. To achieve this target, Infrastructure Ontario (IO) was directed by the MOI to implement building electricity reduction initiatives.

Moving forward, the MOI has directed IO to implement a new target encompassing all fuel types (electricity, natural gas, propane, fuel oil, chilled water, steam). The new target consists of a 2% reduction year over year for all fuel types.

Considerable progress has been made since the Ontario government committed to emission reduction targets, with results being achieved through a variety of initiatives such as major building retrofits, retro-commissioning, as well as new design guidelines and standards. Although the 2014 target was achieved, the 2020 target remains uncertain.

As noted above, in the spring of 2016 the government released its five-year *Climate Change Action Plan* in which the government seeks to reduce GHG emissions from its facilities. A new target was established: a 50% reduction in GHG emissions based on 2006 levels by 2030. IO-managed facilities are heading in the right direction and demonstrating improvements in EE and conservation. With aging buildings, investment in new technologies will improve the chances of achieving emission targets.

In January 2017, IO and the IESO announced they had concluded a five-year partnership to test innovative CDM technologies, processes and program delivery models in the public sector. With a view to accelerating the adoption of CDM solutions in both the public and private sectors, IO will receive up to $10 million in support from the Conservation Fund.

Through this agreement, the IESO will support new CDM pilot projects across the diverse IO portfolio which comprises approximately 4,000 buildings and structures across the province that house a wide range of ministry programs. This includes multi-ministry office buildings, courthouses and correctional facilities.

Approved projects are expected to yield insights in a number of areas: facility benchmarking; building automation; HVAC and advanced controls; lighting systems; building envelopes; EE options for heritage buildings; efficient building operation; electricity, gas and water collaboration; energy storage technologies; energy system integration with regional energy planning; and opportunities to work toward net-zero buildings.

By creating a safe controlled environment in which new technologies and solutions are tested, this partnership is expected to reduce some of the existing barriers to widespread EE implementation, as well as encourage uptake in the broader public sector and potentially the commercial, industrial and institutional sectors.

Financial Mechanisms

For Ontario customers considering EE improvements in their homes or businesses, the primary sources of funding are the IESO conservation programs delivered mainly through LDCs.

To help Ontario customers become more energy aware and enable them to better control energy consumption and costs, the Ministry of Energy has compiled a list of incentive programs for businesses[51] and communities.[52]

Some other options include:

The Atmospheric Fund (TAF): The TAF is a not-for-profit organization that invests in urban solutions to reduce GHG emissions and air pollution. As part of its mission, TAF provides financing to entrepreneurs whose products or services can significantly cut emissions in the Toronto area. Developers and property owners who are making their buildings more energy efficient also qualify. The TAF's ultimate goal of reducing Toronto GHG emissions by 80% by 2050 informs all its actions including investment decisions. TAF's vision is helping Toronto become a climate-smart city that is a leader among urban centers around the world.[53]

Efficiency Capital (EC): This privately-held Canadian corporation offers financing for EE retrofits in multi-unit residential, commercial, industrial and institutional buildings through its insured energy savings warranty program. EC enables property owners and asset managers to increase overall building profits by realizing savings on operating costs and achieving significant reductions in GHG emissions associated with energy consumption. EC earns back its capital and professional services investment by sharing clients' utility cost savings.[54]

City of Toronto Home Energy Loan Program (HELP): HELP is a financing option offered by the City of Toronto that provides low-interest loans to local homeowners interested in improving the energy and water efficiency of their homes. Through HELP, the City provides the funding required to complete improvements, and homeowners repay the loan over time through installments on their property tax bill.[55]

Green Investment Fund (GIF): Created to support the objectives of the Ontario *Climate Change Action Plan*, this $325-million fund supports projects that fight climate change, grow the economy and create jobs. These investments are part of the government plan to secure a healthy, clean and prosperous low-carbon future. GIF projects fulfill the following: help homeowners use less energy; support the installation of more electric vehicle charging stations across Ontario; retrofit social housing developments; help businesses reduce emissions; help fund local environmental organizations; and help provide Indigenous communities with training, tools and infrastructure to address climate change.[56]

IESO funding programs: While not exclusively focused on CDM, the IESO offers a number of funding programs to help different groups participate in the energy sector. The Education and Capacity Building Program is intended to support First Nations and Metis communities and organizations, coops, municipalities, public-sector entities, registered charities and not-for-profit organizations. These projects help equip communities and organizations with knowledge and training, creating opportunities for them take part in the energy economy.[57]

The Aboriginal Community Energy Plan Program helps Indigenous communities find ways to consume less electricity, manage energy costs and explore opportunities for renewable energy solutions.[58]

SMART Green: This program, offered by the Canadian Manufacturers & Exporters in partnership with the Government of Ontario, helps small and medium sized manufacturers reduce GHG emissions and improve EE. It provides assistance in the form of 50% non-repayable grants for eligible costs up to $200,000 and is intended to improve the competitiveness of the Ontario manufacturing supply chain by not only delivering value to eligible manufacturers, but also contributing to Ontario's 2020 GHG emission reduction targets.[59]

High-rise Retrofit Improvement Support Program (Hi-RIS): This is a three-year pilot program offered by the City of Toronto to help residential property owners undertake energy and water efficiency and conservation improvements. The multi-residential stream of the program has a $10-million envelope with a participation target of approximately 10 buildings.[60]

Guelph Energy Efficiency Retrofit Strategy (GEERS): As part of the GEERS program, Guelph homeowners are encouraged to make EE improvements to their homes, including insulation, windows, weatherizing, climate control devices, furnaces, air conditioners, heat recovery systems and solar panels. The upfront costs of the retrofits are paid through the pilot program, and homeowners repay these costs over a period of up to 25 years at low interest rates through special monthly charges known as Local Improvement Charges which are collected through property taxes.[61]

CMHC On-reserve Housing Retrofit Initiative: This federal program offers financial assistance to First Nations for housing renovations and general improvements under existing CMHC agreements. Retrofits to increase EE are eligible costs.[62]

Hydro One First Nations Conservation Program: This program for First Nations communities provides funding for EE upgrades that help save energy and make homes more comfortable.[63]

Hydro One Remotes Program: This portfolio includes a Commercial Lighting Retrofit Program that helps customers make upgrades to lighting systems in existing buildings, a Street Lighting Retrofit Program that helps customers with low or high-pressure sodium streetlights make the transition to LED streetlights, and a Mail-In Rebate Program.

FIMUR—Homeowner Repair Program: This program helps low to moderate-income Indigenous homeowners make eligible home repairs. It is offered to First Nations, Metis or Inuits who own an off-reserve home. It has five funding priorities, one of which is EE, and the program runs annually until 31 March 2020.[64]

SECTION II—NATURAL GAS SECTOR

Background

Natural gas conservation efforts in Ontario date back nearly 25 years. The province counts close to 3.5 million natural gas customers. [65] The initial regulatory framework for natural gas DSM was established in 1993 through a report[66] of the OEB which regulates the province's electricity and natural gas sectors.

The OEB has been promoting DSM and approving DSM plans for the major gas utilities ever since. Although Ontario has three rate-regulated gas distributors—Enbridge Gas Distribution Inc. (Enbridge), Union Gas Limited (Union) and Natural Resource Gas Limited—only the two largest utilities, Enbridge and Union, file regular DSM plans. Two municipally-owned gas companies (City of Kitchener and City of Kingston) are not rate-regulated by the OEB.

The aforementioned OEB initial report included implementation guidelines which remained effective until 2006 when a generic proceeding[67] was undertaken to address a number of common issues related to natural gas distributor DSM activities. After consultation with the sector, changes were applied for the 2007-2009 period, after which the OEB reviewed the DSM Framework and issued the 2012 DSM Guidelines in June 2011.[68]

The 2012 DSM Guidelines took into consideration the *Green Energy and Green Economy Act, 2009,* which was enacted as the guidelines were being developed.[69] The guidelines provided the gas companies with general policy direction and guidance and served as the foundation for their DSM plans from 2012 to 2014.

The OEB subsequently received a directive[70] from the Minister of Energy in March 2014, entitled the Conservation Directive, which required the OEB to develop a new six-year DSM framework that met specific government objectives including greater coordination and alignment between the gas and electricity sectors, among other things. In addition to including policies related to key elements to be funded through gas utility distribution rates, the framework provided guidance to enable the gas utilities to develop their 2015 to 2020 DSM plans.[71]

As part of the 2015 to 2020 DSM Framework, the OEB recognized that DSM offers many benefits. The OEB, however, stated that approval to fund such programs through distribution rates must be within the scope of its legislative mandate. The OEB further clarified that ratepayer-funded DSM programs should focus on the following goals:[72]

1. Assist consumers in managing their energy bills by reducing natural gas consumption;

2. Promote energy conservation and EE to create a culture of conservation;

3. Avoid costs related to future natural gas infrastructure investment, and improve the load factor of natural gas systems.

Ontario's rate-regulated gas utilities were provided guidance on developing multi-year DSM plans that include annual performance targets and longer-term goals for their DSM programs. They are required to include proposed conservation targets as part of their applications for distribution rates to fund their DSM plans. They are expected to develop annual and long-term DSM budgets that reflect the costs of administering and delivering DSM programs. Program activities include marketing and education for consumers, as well as financial incentives to program participants. To help the utilities prepare their plans and projections, the OEB provides a budget cap to each utility on an annual basis.

Under this framework, gas utilities have the option of adhering to established guidelines and/or proposing alternatives in their plans. If they choose to submit an alternative, they must present the evidence and rationale that support the proposal and clearly demonstrate how the public interest is enhanced.

Once the gas utilities submit their multi-year DSM plans for approval, the OEB reviews and assesses the appropriateness of the proposed budget amounts relative to the natural gas savings targets (both annual and long-term goals). The OEB analysis also includes the resulting impact that planned DSM-related spending is expected to have on distribution rates.[73]

Ultimately, distribution customers pay for all natural gas DSM activities in Ontario. Although some customers participate in the programs offered by the gas utilities and benefit directly from reducing their consumption, others do not. The OEB considers the overall impact that DSM costs have on all customers (both participants and non-participants) when it evaluates DSM plans and distribution rate applications.

Enbridge and Union filed an application before the OEB for approval of their 2015 to 2020 DSM plans under Section 36 of the Ontario Energy Board Act, 1998. Approved DSM budgets are recovered from customers through distribution rates. As part of the OEB 2015-2020 DSM Decision on Enbridge and Union's respective applications, rendered on January 20, 2016, the OEB stated that a mid-term review would serve to assess performance on annual metrics, budget levels, impacts on customer rates, and shareholder incentives. The OEB also outlined a number of studies and reports that Union and Enbridge would be required to submit as part of the mid-term review. The OEB released a letter on June 20, 2017, initiating the mid-term review and providing details on the scope, approach, and schedule thereof. Tables 10-3 and 10-4 present the DSM programs approved by the OEB, as well as the stakeholders involved and their respective key responsibilities.

Table 10-3: Ontario Natural Gas DSM Programs Approved by the OEB

Residential Programs, Including Low Income	
Enbridge	**Union**
Home Energy Conservation	Home Energy Rebate
Residential Savings by Design	Optimum Home Program
Home Winterproofing (low income)	Home Weatherization (low income)
	Aboriginal
	Furnace End of Life Upgrade
Low-Income Multi-Residential Affordable Housing	Low-Income Multi-Family
Low-Income New Construction	
Business and Industrial Programs	
Commercial & Industrial Prescriptive	Commercial & Industrial Prescriptive
Commercial & Industrial Direct Install	Commercial & Industrial Direct Install
Commercial & Industrial Custom	Commercial & Industrial Custom
Small Commercial New Construction (pilot)	
Commercial Savings by Design	Commercial Savings by Design
Run it Right	RunSmart
Comprehensive Energy Management	Strategic Energy Management
	Large Volume
Other Programs	
Energy Literacy	
School Energy Competition	

Stakeholders

Table 10-4: Natural Gas DSM Stakeholders and Their Key Responsibilities

Organization/Group	Key DSM-Oriented Responsibilities
Ministry of Energy (MOE)	• Establishes the policy framework through the province's Long-Term Energy Plan • Sets energy and water efficiency standards for appliances and products • Demonstrates conservation leadership
Ministry of the Environment and Climate Change (MOECC)	• Establishes the policy framework for reducing the impacts of climate change through the Climate Change Action Plan • Oversees the cap-and-trade program • Works with the IESO on programs offered through the Green Ontario Fund, including Whole Home
Ontario Energy Board (OEB)	• Regulates the electricity and gas sectors • Approves gas utility distribution rates • Regulates gas DSM and approves gas utility DSM plans • Develops DSM policy framework in accordance with government direction
Independent Electricity System Operator (IESO)	• Collaborates with gas utilities on different operational issues, as well as certain aspects of the design and development of the Whole Home program • Works with MOE and MOECC on programs offered through the Green Ontario Fund, including Whole Home • Operates the province-wide electrical grid to ensure supply and demand are balanced • Manages the CDM electricity portfolio
Local Distribution Companies (Gas)	• Work directly with customers on DSM projects • Work with the IESO on certain aspects of the design and development of the Whole Home program
Channel Partners	• Cross sell DSM programs with other products • Provide energy services to customers • Design and deliver DSM programs
Environmental Commissioner Office	• Submits an annual Energy Conservation Report to the Legislative Assembly of Ontario, including progress to reduce GHG emissions
Municipal Governments	• Develop corporate and community energy plans • Provide input on regional energy plans • Develop and implement policy • Demonstrate conservation leadership
Canadian Standards Association	• Develops EE standards on products
Natural Resources Canada (NRCan)	• Administers federal Energy Efficiency Regulations which establish EE standards for energy-using products • Oversees the ENERGY STAR® initiative, a voluntary partnership between the Government of Canada and industry to promote EE
Indigenous Communities	• Develop community energy plans • Provide input on regional plans

Policies, Strategies and Regulatory Framework
Natural Gas and Climate Change

As described in the Electricity section of this chapter, the Legislative Assembly of Ontario passed the *Green Energy and Green Economy Act*, 2009, to encourage the development of renewable energy generation and promote energy conservation. The adoption of the *Climate Change Action Plan* seven years later was another major milestone in transitioning Ontario to a low-carbon economy.

In May 2016, the province finalized the rules for its cap-and-trade program which is intended to limit GHG emissions, reward innovative companies, generate opportunities for investment in Ontario and create jobs.[74] The cap-and-trade regulation took effect on July 1 of that year. The *Climate Change Action Plan* and the cap-and-trade program are key elements of Ontario's strategy to reduce GHG pollution to 15% below 1990 levels by 2020, 37% by 2030 and 80% by 2050. The government will report on plan implementation annually and renew the plan every five years.[75]

As is the case with electricity, natural gas is a priority area for energy planning in Ontario, which falls under the mandate of the Ministry of Energy which is legislatively responsible for producing a Long-Term Energy Plan (LTEP). The latest iteration was released in the fall of 2017. Gas utilities are in charge of their own gas supply planning and obtain approval from the OEB.

The 2013 LTEP[76] and the 2014 Conservation Directive stress the importance of aligning natural gas DSM efforts with electricity CDM efforts to support the government's overarching Conservation First mandate.

Gas DSM Funding

Natural gas DSM in Ontario is funded differently than electricity CDM, with all funds being recovered from natural gas customers through distribution rates which are approved by the OEB. The OEB is required to protect the interests of consumers with respect to prices, reliability and quality of gas services while promoting energy conservation and EE. As noted, the OEB must consider the rate impact on customers when it assesses funding levels for gas DSM.

With these requirements in mind, the OEB confirmed[77] that for DSM activities between 2015 and 2020, gas utility annual DSM budgets should be guided by the principle that DSM costs (including both DSM budget amounts and shareholder incentive amounts) for a typical gas utility residential customer should be no greater than $2.00/month—a substantial increase over the previous framework which capped costs at $1.00/month.

Another fundamental difference between the electricity and gas sectors is that unlike electricity distributors, natural gas utilities are not licensed by the OEB. Instead, they operate under franchise agreements with the municipalities they serve. In practical terms, this means there is no license condition mandating that gas utilities undertake DSM activities. These activities are voluntary. To motivate gas utilities to actively and efficiently pursue DSM savings and to recognize exceptional performance, the OEB makes a shareholder incentive available, though it is not tied to gas utility DSM budgets.[78] For 2015-2020, Enbridge and Union will potentially benefit from a total annual maximum of $10.45 million each if performance targets are met.

Evaluation, Measurement and Verification

As is the case with electricity conservation, the EM&V of energy savings is a critical component of the natural gas DSM framework. In addition to enabling impact assessments (e.g. reduced consumption) and the effectiveness of EE programs, EM&V is also used to identify program improvements that offer the potential to drive stronger performance.

For the 2015-2020 DSM Framework, the OEB chose to adopt a larger role in the program evaluation process, a role previously managed by the gas utilities with input from key stakeholders. In addition to conducting annual evaluations on program results, which are published every year, the OEB has indicated that it will conduct multi-year impact assessments of selective gas utility DSM programs on a periodic basis (i.e. every three years).[79]

To align natural gas DSM programs with electricity CDM programs and ensure government objectives are considered, the OEB concluded that the same cost-effectiveness approach should be used for screening natural gas DSM programs as electricity CDM programs. [80] The OEB also indicated it would adopt an enhanced TRC test, or the TRC-Plus test, which gas utilities were instructed to use to screen all potential DSM programs when developing their multi-year DSM plans. Gas utilities were also directed to incorporate the PAC test as a secondary reference tool to calculate cost effectiveness and screen potential DSM programs.

Gas DSM and the Public Sector

As previously discussed in the above Electricity section, Ontario Regulation (O. Reg.) 397/11[81] has transformed the way public agencies manage their energy consumption. This regulation, which falls under the *Green Energy Act, 2009*, came into force on 1 January 2012 and requires broader public-sector organizations, including hospitals, municipalities, universities, colleges, school boards, as well as municipal service boards responsible for water and sewage treatment and pumping operations, to report annually to the Ministry of Energy on their energy use and GHG emissions. These reports must be published on their respective websites, effective 1 July 2013. In addition as of 2014, public agencies have been required to develop five-year conservation plans and publish these on their websites. Plans must be updated every five years beginning in 2019.

The Ministry of Energy claims energy reporting and conservation planning will help public agencies manage electricity use and costs, identify best practices and energy-savings opportunities, evaluate results by comparing to similar facilities across the province, help set goals by providing a benchmark, and measure improvements over time. [82]

As noted above, municipalities are among the broader public-sector entities required to comply with O. Reg. 397/11. The Government of Ontario supports local energy planning through the Municipal Energy Plan (MEP) program. Introduced in 2013, the MEP program enables municipalities to understand their local energy needs, identify EE and clean energy opportunities, and develop plans to meet their goals.[83]

Financial Mechanisms

Enbridge and Union both offer DSM programs for residential, commercial and industrial customers. While some overlap exists between the two portfolios, each utility designs and delivers programs specific to its own customer base, which include training, tools, rebates and incentives.

Chapter 11

Prince Edward Island

ENERGY SECTOR

The energy consumption of Prince Edward Island increased slightly between 1995 and 2009 before peaking in the 2010-2012 period at an annual average of 626 ktoe. Consumption dropped in the 2013-15 period (578 ktoe/year) compared to 2004-2009 levels, nonetheless representing an 11% increase compared to 1995-97 levels.

As of 1995-97, industrial sector consumption constantly increased, effectively more than doubling (130%) in the 20-year period from 1995 to 2015. In 2013-15, this sector consumed nearly 20% of total energy consumption in the province. The transportation sector consumed the most with energy consumption increasing twice as rapidly as the total increase in energy consumption over the 20-year period (22%). This sector therefore accounted for 42% of 2013-15 energy consumption. Both the residential and commercial sectors decreased building energy consumption in the 20-year period by 14 and 32% respectively. Throughout the 2013-15 period, the residential sector consumed 18% of total consumption, while the commercial sector consumed 11%. Energy consumption in the agriculture sector increased by 16% between the 1995 and 2015 periods. This sector consumed slightly more than 8% of total energy consumption in 2013-15, while for this same period the public administration sector accounted for nearly 2% of total provincial consumption.

Figure 11-1 illustrates the average energy consumption in three-year periods from 1995 to 2015.

ENERGY EFFICIENCY BACKGROUND

Prince Edward Island (PEI) is the smallest Canadian province in terms of land area. Along with Nova Scotia and New Brunswick, it is one of three Maritime Provinces located on Canada's East Coast. The

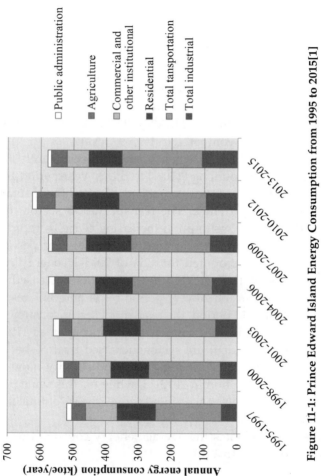

Figure 11-1: Prince Edward Island Energy Consumption from 1995 to 2015[1]

province is comprised of the main island of the same name and several much smaller islands. Given its insular and isolated location, the province's energy sector is unique in Canada. PEI is indeed highly dependent on imported energy sources and fossil fuels, a situation that makes the province highly vulnerable to energy price fluctuations. Figure 11-2 illustrates the energy mix (2014), 67% of which is imported petroleum.

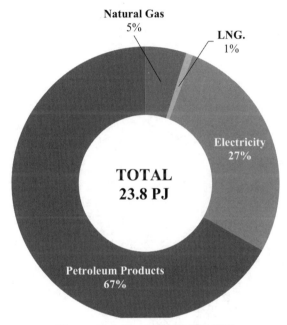

Figure 11-2: Energy Mix (2014)[2]

The advantages of EE for the province are thus threefold since it benefits the local economy while reducing GHG emissions and reliance on imported fossil fuels. Although different initiatives undertaken during the last decade have successfully fostered the development of renewable energy projects with a significant increase in wind electricity production, the effectiveness of EE initiatives has been rather limited. However with the release of its new ten-year Energy Strategy in March 2017, the government placed EE and energy conservation at the heart of economic development. The timeline of EE initiatives in PEI is illustrated in Figure 11-3.

PEI became one of the first provinces to sign a Climate Change Memorandum of Understanding (MOU) with the Government of Cana-

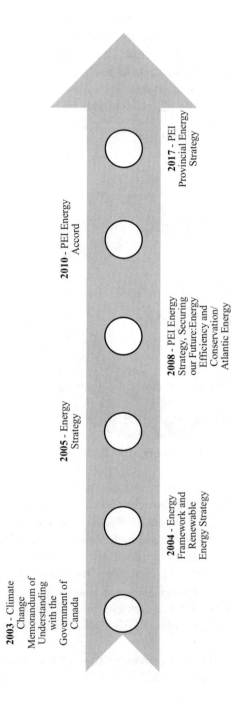

2003 - Climate Change Memorandum of Understanding with the Government of Canada

2004 - Energy Framework and Renewable Energy Strategy

2005 - Energy Strategy

2008 - PEI Energy Strategy, Securing our Future:Energy Efficiency and Conservation/ Atlantic Energy Framework for Collaboration / establishment of PEI Office of EE

2010 - PEI Energy Accord

2017 - PEI Provincial Energy Strategy

Figure 11-3: Timeline of EE Initiatives in PEI

da in 2003. The MOU generally established the commitment of both parties to coping with climate change. In 2004, the first Energy Framework and Renewable Energy Strategy was released, which outlined 19 actions for the provincial government, including among other things:

- Publication of the Renewable Energy Act;

- Establishment of a Climate Change Committee to consult with Islanders;

- Promotion of wind and hydrogen technology through a Hydrogen Village™ Concept;

- Comprehensive PEI Energy Strategy released in the spring of 2005;

- Presence at Atlantic Energy Ministers' Forum.[3]

2008 was an important year for the EE sector in PEI because the second version of the Energy Framework and Renewable Energy Strategy was released. The document stated that almost all actions from the previous strategy were completed and the new version would expand on these actions to cope with issues in the energy sector. This strategy helped the PEI Government establish its role in promoting EE and energy conservation, the use of renewable energy and biofuels, as well as awareness-raising activities. The strategy specifically focused government efforts in three areas: energy security, environmental sustainability and economic development.[4]

In 2007, the government Energy and Minerals Division hired an expert to evaluate the savings potential of current EE initiatives and their anticipated impacts up to 2017. The results were published in a report which initiated the establishment of the PEI Office of Energy Efficiency (OEE) in 2008. The office was in charge of fostering consumer awareness about EE and implementing EE programs. The OEE has commissioned more than 7,000 energy audits and inspections of PEI homes, which led to financial assistance in the form of loans and grants for residential EE improvements.[5]

In 2008, the governments of PEI, New Brunswick, Newfoundland and Labrador, and Nova Scotia signed the Atlantic Energy Framework for Collaboration to foster cooperation on sustainable development, as well as ensure secure and reliable energy supply for these provinces. In 2010, the Government of PEI concluded the PEI Energy Accord with the Maritime Electric Company Limited to stabilize energy prices and increase PEI reliance on locally owned wind power.

The most recent version of the energy strategy (PEI Provincial Energy Strategy, 2016/17) was released in 2017. Similarly to previous strategies, it was preceded by open public discussions. The new ten-year strategy established goals in the following sectors:

- Energy efficiency and energy conservation;
- Power generation and management;
- Biomass and heating;
- Transportation.[6]

PEI remains the only province without a building code, although codes exist for the cities of Charlottetown, Summerside and Stratford.

STAKEHOLDERS[7]

The Government of PEI manages energy policy and energy-related issues through the PEI Energy Corporation, as well as through the Energy and Minerals Division of the Department of Transportation, Infrastructure and Energy.

The Energy and Minerals Division is in charge of developing, implementing and administering energy policies, programs, mineral resources and gas exploration initiatives in PEI. The Energy and Minerals Division also manages the operations of the OEE.

Efficiency PEI, known as the OEE before 2016, was established in 2008 as a division of the Department of Transportation, Infrastructure and Energy to assist Islanders in reducing their energy consumption and minimizing the environmental footprint of daily activities. The mandate of Efficiency PEI is to provide citizens with advice and programs that promote sustainable energy use and reinforce the importance of sound energy management for the economic, social and environmental well-being of residents and businesses.[8]

Since its inception, Efficiency PEI actions have mostly focused on the residential sector with a few programs also targeting the commercial and institutional sectors. To date, six EE programs are available to Islanders, including three rebate programs for the installation of retrofit equipment, two audit programs, and one Home Energy Low-Income Program (HELP).[9] While Efficiency PEI has been responsible for non-electric EE, the task of leading electricity efficiency initiatives has been left to Maritime Electric.

PEI Energy Corporation, a Crown Corporation, was established in 1978 with the main objective of coordinating all energy-related issues. It supports local industries in the development of new technologies and provides financial assistance to develop and install new energy systems to achieve diversity of supply and environmental sustainability. Maritime Electric and Summerside Electric are the two main electric utilities of the province. While the former has carried out some EE activities for its customers pursuant to the Electric Power Act, the latter has not been as active.

Maritime Electric is the main distribution and generation utility in PEI. The utility sources electricity through third-party supply contracts and from its own power generation installations. The submarine cable interconnection with the mainland is an important component of the province's electricity supply system, without which the Island would be highly reliant on oil. Thanks to this interconnection, PEI can purchase electricity from nearby New Brunswick, Nova Scotia, Quebec, and New England. The Maritime Electric electricity generation installations consist of two plants, the Charlottetown Thermal Generating Station (Charlottetown Thermal Plant), and the General Electric LM6000 simple-cycle combustion turbine generator. Maritime Electric offers tools on its website to evaluate the energy consumption of homes and businesses and tips to reduce energy use. The current core power purchase agreement of Maritime Electric is with NB Power, a relatively large utility that operates 14 generating stations. More than half of PEI's electricity is distributed by NB Power.

Summerside Electric's generation assets consist of the Harvard Street Station (an oil-fired plant) and the City of Summerside Wind Farm. It maintains power purchase agreements with both IPR-GDF Suez North America and NB Power. Summerside Electric also operates Heat For Less Now!, an electric thermal storage program.

The Island Regulatory and Appeals Commission (IRAC) regulates the supply of energy, as well as the overall performance and future planning activities of PEI utilities. The IRAC chair and two commissioners are responsible for electricity regulation in PEI. They undertake regular activities and hold periodic hearings related to Maritime Electric, including reviews of the capital budget, rate applications, monthly financials, and investigations pursuant to customer complaints.

POLICIES AND STRATEGIES

The PEI government manages energy through the Department of Transportation, Infrastructure and Energy which adopted the Prince Edward Island Energy Strategy 2017.[10] The previous versions of the strategy were released in 2004 and 2008, all of whose goals were accomplished. The new version expands future actions based on past successes.

The PEI Energy Strategy 2017 is a ten-year strategy covering EE, energy conservation and renewable energy sources based on three main principles: reducing GHG emissions; implementing cost-effective measures; and developing local economies.

The first and very important step in the strategy is reducing energy consumption. It outlines the main EE and energy conservation targets of achieving a 2% reduction in electricity and energy use per year by 2020 and sets up an independent utility with the mandate of pursuing EE. To achieve these, the government is looking at implementing more programs to encourage Islanders to make their homes more efficient and tailoring the programs to include all fuel sources. The government wants to make it easier for Islanders to access these programs which cover almost all sectors: public, commercial and industrial. The objectives of the Energy Strategy 2017 are as follows:

- Residential: Implement a set of EE programs that enable customers to cost-effectively reduce energy consumption; implement a low-income residential program accessible by all PEI residents. Programs will include activities to foster appliance recycling and the sale of efficient appliances (through rebates), as well as deep energy retrofits of building envelopes and the construction of new buildings.

- Commercial, industrial and institutional: Implement a set of programs which include small business programs that encourage selling the most efficient appliances and lighting, as well as custom purchase options for large customers.

- Geo-targeted EE: Develop sets of geo-targeted EE protocols and guidelines for transmission and distribution grid maintenance.

- Codes and standards: Implement the National Building Code and National Energy Code for Buildings across the province; develop and adopt a provincial building code with EE level requirements above the National Building Code; support pilot projects for high efficiency buildings; enforce mandatory building labeling disclosure in the residential sector, which involves conducting home energy audits and issuing an energy performance score (this process ensures that the potential buyer has information about the energy costs of the home); examine the feasibility of a mandatory commercial/institutional energy reporting system for buildings; monitor appliance standards approved in the United States or in other Canadian jurisdictions.

- Demand response: Develop strategies for the residential and SME sectors to reduce electricity demand during peak hours; encourage large commercial and institutional building operators and owners to install energy management systems and enroll in demand response programs.

LEGAL AND REGULATORY FRAMEWORK

The PEI government has enacted no EE act. The Electric Power Act, adopted in 2004, is the primary legislation used to regulate the supply and delivery of electricity in the province. It states that IRAC is the regulatory agency for electricity rates and utilities and carries out regulatory oversight of DSM activities. The first DSM plan in PEI was adopted in 2004 under the Renewable Energy Act. The objective of the Act was to reduce the intensity of peak electricity demand for utilities. The ensuing DSM programs fell under the purview of Maritime Electric. The utility is equipped to lead DSM programs. It has the expertise, direct access to relevant system information and the working knowledge necessary to design and manage DSM initiatives.[11]

The PEI government has not adopted building or equipment codes and standards. Only three cities enforce the National Building Code, namely Charlottetown, Summerside and Stratford, while construction companies outside these cities are not required to respect the code. Nonetheless, the Energy Strategy 2017 establishes some targets in this area.

The government has not yet adopted a carbon tax or cap-and-trade system, but this issue is currently being discussed.

FINANCIAL MECHANISMS

According to the NRCan website, six programs are currently accessible in PEI, as follows:
- Building Envelope Upgrade Rebate;
- Heat Pump Rebate;
- Equipment Upgrade Rebate;
- Home Energy Audit Program;
- Home Energy Low-Income Program (HELP);
- Commercial Energy Audit Program.

All programs are delivered by Efficiency PEI.

The Building Envelope Upgrade Rebate program offers rebates for the installation of insulation, replacement windows and doors, and air sealing improvements in the residential sector.

The Heat Pump Rebate program provides rebates to residential users for ENERGY STAR® Most Efficient air source and ground source (or geothermal) heat pumps.

Another residential program, Equipment Upgrade Rebate, offers rebates for the installation of water saving devices, biomass heating devices, other energy-saving products and ENERGY STAR® certified heating equipment.

The residential sector Home Energy Audit Program partly subsidizes ($150) the cost of an EnerGuide home evaluation. In the event that homeowners choose to proceed with efficiency upgrades after the audit, Efficiency PEI further subsidizes costs by enhancing the corresponding rebates for which the home qualifies.

HELP is intended for low-income residential customers. It provides free kits that include air sealing, energy efficient light bulbs, low-flow showerheads, programmable thermostats, and a voucher for a free furnace/boiler cleaning.

The only program that targets the commercial and public sectors is the Commercial Energy Audit Program. It offers financial incentives of up to $1,000 for an evaluation to determine the potential for EE upgrades in commercial buildings.[12]

Chapter 12

Quebec

The Province of Quebec has long positioned itself as one of the most prolific jurisdictions in the production of clean energy thanks to its abundant supply of water which has been harnessed to generate electricity. While leadership in the field of renewable energy has earned the province international recognition, it has paradoxically dampened the urge to reduce the province's energy consumption and develop its EE market. The electricity consumption of Quebecers is among the highest in both Canada and the world. Although this is partly explained by the presence of highly energy intensive industries attracted by the low cost of electricity, this same reason has long rendered most EE initiatives ineffective as a result of lack of interest among stakeholders. Nonetheless, Quebec has a long history with and much experience in EE which has been included in energy strategies to reduce energy waste. Today, EE is recognized as a tool to spur the economy, achieve environmental targets, and create wealth.

ENERGY EFFICIENCY BACKGROUND

Similarly to other Canadian provinces and international jurisdictions, the concept of EE in Quebec took root in the wake of the two oil crises of the 1970s that undermined the energy sector. In this context, the first governmental entity dedicated to EE matters was established in 1977, the Bureau des économies d'énergie (Energy Savings Bureau). The Bureau's mandate was to assess energy savings potential throughout the province and propose government policies and programs aimed at reducing energy consumption. Amongst the main actions undertaken by the Bureau was the implementation of a home insulation program and the promotion of energy savings in the public and parapublic sectors.[1] In 1988, the Bureau changed its name to the Bureau de l'efficacité énergétique (BEE, Energy Efficiency Bureau). Although minor, this

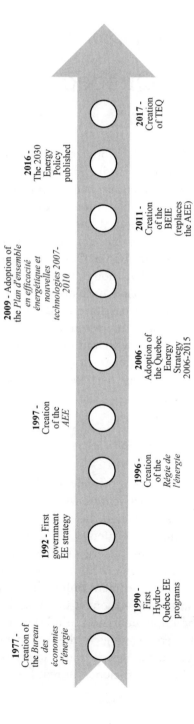

Figure 12-1: Timeline of Main Energy Efficiency Events in Quebec

name change exemplifies the paradigm shift that occurred at the time
in the way energy savings were perceived. Whereas the term energy
savings was mostly used to designate a reduction in energy waste, the
term energy efficiency was coined to designate the integrated approach
of using energy to simultaneously reduce consumption, strengthen the
economy, and support sustainable development.

QUEBEC, BIRTHPLACE OF THE FIRST ESCO IN CANADA[2]

The Province of Quebec is a pioneer in EE and energy performance contracting (EPC). In 1981, it established the first Canadian ESCO and one of
the first in the world, namely Econoler Inc. The firm, established as a joint
venture between the province's public electricity utility Hydro-Québec
and a private-sector engineering firm, developed a new concept based on
a unique shared savings approach with a first out option (an open book
approach whereby contracts are terminated upon complete payment of all
project costs even if the contract period has not come to term). The market
was very attracted by this model. Through EPC, Econoler aimed to develop and implement EE projects by offering turnkey services enhanced by an
integrated financing offer and an investment payback method solely based
on savings generated by projects. This marked the beginning of what we
refer to today as the EPC concept.

Between 1981 and 1989, more than 1,000 projects were implemented
based on Econoler's unique concept in all types of commercial, institutional and industrial establishments across Quebec and Canada. The firm directly invested over $135 million to implement these projects, while several
clients chose to finance them on their own, though always under the EPC
approach. These projects generated yearly recurring savings of $35 million,
and most were recognized as extremely successful from both a technological and commercial perspective.

As early as 1983, the rest of Canada began expressing interest in the
new concept pursuant to Econoler's success in Quebec. Canertech, a subsidiary of Petro-Canada, sought to replicate the Econoler model in Quebec
and then apply this expertise elsewhere in the country. Econoler therefore
signed a licensing agreement with Canertech which then transferred the
newly acquired knowledge and skills to its newly created subsidiaries in
Ontario, New Brunswick, Nova Scotia, and Prince Edward Island. Starting
in 1985, other companies were founded based on the same model, which
subsequently fostered competition in the Canadian market. Thus, the Canadian ESCO industry was born.

In 1992, this new approach was adopted under the first EE strategy of the province (La Stratégie Québécoise d'efficacité énergétique: une contribution au développement durable), which provided a framework with objectives, intended actions, and means of intervention to promote and facilitate stakeholder efforts in using energy more efficiently. More specifically, this new strategy was intended to reduce energy intensity in the Quebec economy by 15% by 2001. The role, mandate, and structure of the BEE, renamed the Office of Energy Efficiency (Direction de l'efficacité énergétique, DEE) in 1994, were also revised and the office was placed under the authority of the Minister of Natural Resources and Energy (Ministère des Ressources naturelles et de l'Énergie, MERN). This restructuring aimed to align the DEE mission with climate change issues about which governments could no longer remain idle. With major involvement of Hydro-Québec staff in the development of new EE programs, the vocation of the DEE became more orientated toward elaborating policies and collaborating with major actors in the sector rather than implementing programs.

By the early 1990s, Hydro-Québec became a leading institutional force in the EE market by investing heavily through a massive grant program. The objective was to use EE as a competitive source of energy to target both energy consumption reductions in all economic sectors as well as address peak demand issues at the utility level. Over $300 million was invested over a period of five years in the first half of the 1990s with the creation of approximately 15 DSM programs. These programs included various kinds of initiatives:

- Market transformation activities:
  - Educating and raising awareness among clients;
  - Training market participants;
  - Stimulating EE technology offerings through research and development and by influencing technology providers;
  - Contributing to regulatory changes.

- Concrete activities with short-term impacts:
  - Direct installation of equipment such as electronic thermostats;
  - Incentives and rebates for the purchase and installation of EE technologies.

These initiatives ultimately allowed Hydro-Québec to generate nearly 2.5 TWh in electricity savings per year and reduce yearly peak demand by approximately 400 MW.[3]

Building on this momentum in 1996, the government published a new energy policy entitled Energy Serving Quebec: A Sustainable Development Perspective (L'énergie au service du Québec: Une perspective de développement durable) which further emphasized the importance of EE to support sustainable economic development and increase the availability of the province's energy resources. One of the key measures of this new policy was the creation of the Energy Efficiency Agency (l'Agence de l'efficacité énergétique, AEE), which replaced the DEE. Although the role of this agency remained very similar to its predecessors, it had by then become an independent agency tasked with new responsibilities such as assisting the newly created Energy Bureau (Régie de l'énergie) on EE matters and promoting EE in all sectors. The AEE was also mandated to design and implement programs to support new initiatives in the EE field.[4] Some of the then created programs are still active today, namely Novoclimat, a financial aid and certification program for new energy efficient home constructions, and Éconologis which offers EE services to low-income households.

Between 1996 and 2006, the AEE developed and consolidated its leadership position as one of the most exemplary public institutions on EE matters in Canada. The agency undertook several initiatives to promote EE for all kinds of energy sources in every sector of activity and in all Quebec regions. These services were broken down into four main categories:[5]

- Education and awareness activities:
  - Development of different information tools and training material related to all aspects of EE as a function of clientele needs: a watch center, training manuals, publications, websites, exhibitions, conferences, seminars, etc.

- Demonstration projects:
  - Projects carried out in collaboration with partners to demonstrate the benefits of a given technology, new approach or application that offered interesting EE potential that could be replicated.

- Incentive measures for EE project implementation:

- — Design, management and verification of action plans, and corresponding intervention tools to reach energy savings targets.
- • Counseling and advisory services:
  - — Comments and notices formulated in response to questions related to EE, as well as laws, regulations and standards thereof, from the government or other public organizations.

Among the sectors targeted by the AEE, the residential and institutional sectors by far benefited the most from these initiatives. By 2006, the AEE and its partners had performed more than 25,000 energy evaluation visits of existing households and 2,200 new houses were certified through the Novoclimat program. In the institutional sector of the early 2000s, more than 500 buildings representing approximately 11% of the education and health building stock benefited from funding for feasibility studies to help building operators identify energy savings opportunities. Partners such as Énergir,[i] Gazifère (natural gas utilities) and the Association of Institutional Real Estate Managers (Association des gestionnaires de parcs immobiliers, AGPI) also contributed to efforts in this sector through education and awareness raising in the form of informative meetings with building managers. These partners further helped by disseminating technical factsheets to help in the decision-making process related to EE project implementation. As a result of all the initiatives conducted between 1996 and 2006, the AEE reported total annual energy savings of 1,050 GWh in all sectors.

In 2006, the adoption of the Quebec Energy Strategy 2006-2015: Using Energy to Build the Quebec of Tomorrow (la Stratégie énergétique du Québec 2006-2015: Pour construire le Québec de demain) marked a new era for the energy sector of the province. Energy savings targets were established for all energy sources for the first time, including petroleum products, with the main objective of increasing the overall EE target by a factor of eight compared to then current targets. While EE remained at the core of the new strategy, the development of renewable energy was emphasized, mostly wind turbines and new energy technologies that had emerged as innovative alternatives. To facilitate new initiatives, the AEE became a more autonomous body, and its mandate was broadened to foster and promote innovation in the energy sector. By the end of 2006, the

---

[i] Previously known as Gaz-Métro (Gaz Métro changed its name to Énergir in November 2017).

government had modified certain dispositions in the law that regulates the AEE (Law 46) to more effectively enable the orientations established by the 2006-2015 Strategy. These dispositions entrusted the agency with new responsibilities including the development, implementation, and evaluation of triennial action plans on EE and new technologies. These plans were thus aimed at outlining all EE efforts in which main stakeholders could engage to fulfill government objectives. To that end, electricity and gas providers were required to establish their own triennial targets, intended actions, and programs to be submitted for AEE approval. The AEE, on the other hand, was left in charge of other activities such as EE program content, fossil fuel reduction activities as well as the promotion and support of new energy technologies.

To further support the AEE mission, new dispositions were added to the law in 2008, which provided for a new source of funding from annual duties payable by energy distributors. Pursuant to the adoption of this regulation, the financial contributions to the AEE almost doubled from $35 million for the fiscal year 2007-2008[6] to $67 million in 2008-2009[7]. New programs including four specific to the transport sector were thus developed to address fossil fuel consumption to a greater extent.

Never before had the agency been entrusted with so much responsibility and autonomy, nor had it ever had such vast financial capability to take action. As a result, considerable EE progress was made in all sectors and, by 2009, results surpassed fiscal year targets by more than 12% with total annual energy reductions of 5,077,738 GJ.

The efforts of the AEE, however, were short-lived since the government decided to abolish it in 2011 and reassign its mandates to the MERN by establishing the Bureau of Energy Efficiency and Innovation (Bureau de l'efficacité et de l'innovation énergétiques—BEIE). This new bureau maintained the same funding mechanism based both on annual duties paid by energy distributors and the Green Fund (see rubric below).

STAKEHOLDERS

Ministry of Energy and Natural Resources (MERN) and the Bureau of Energy Efficiency and Innovation (BEIE)

Since 2011, the Act Respecting Energy Efficiency and Innovation has attributed the responsibility of fostering and promoting EE and innovation to the Minister of Energy and Natural Resources through the

THE QUEBEC GREEN FUND[8]

The Green Fund was created in 2006 under the Act Respecting the Ministry of Sustainable Development, Environment and Parks to support Quebec's sustainable development by protecting the environment and preserving biodiversity, as well as fostering the fight against climate change. The Fund derives most of its income from three sources: the sale of emission allowances in the cap-and-trade system; charges payable for the disposal of residual materials; and charges payable for the use of water. The funds generated serve mainly to finance the actions presented in the 2013-2020 Climate Change Action Plan and achieve the GHG emission reduction target of 20% below 1990 levels by 2020.

BEIE. Between 2011 and 2016, the BEIE was thus in charge of elaborating and implementing EE programs or measures aimed at GHG emission reductions and providing technical support for research and development in the field.

Apart from administering EE programs, the Bureau was in charge of supporting research and development in efficiency and innovation, as well as advising businesses and the public on these matters. Various financial incentives were made available to individuals, companies, and industries willing to reduce energy consumption or invest in EE projects. As a result, nine different programs covering every sector and all energy sources were offered by the BEIE, as listed in Table 12-1.

Table 12-1: Portfolio of Quebec Government EE Programs[9]

Residential	Commercial & Industrial	Innovation
• Éconologis	• Drive Green	• Technoclimat
• Heating with Green Power	• Écoperformance	
• Novoclimat 2.0	• Residual forestry biomass	
• Rénoclimat	• Branché au travail (EV workplace charging stations)	

In 2016, with the publication of the new ambitious energy policy (Energy in Quebec: A Source of Growth) and GHG emission reduction targets (see the Policies and Strategies section further below), the Government of Quebec wanted to provide new momentum to the energy sector whose efforts under the BEIE had begun to slow down and stagnate. In April 2017, all BEIE activities and programs were therefore transferred to a new and more autonomous governmental entity named Transition énergétique Québec (TÉQ).

Transition Énergétique Québec (Quebec Energy Transition)
While the government's EE and energy substitution programs have always been the responsibility of and administered by different bodies and thus administratively decentralized, the new 2016 policy included plans to consolidate all these services under a readily accessible entity to improve the experience of program beneficiaries, as well as achieve better coherence and broader synergy between various EE initiatives. The creation of TÉQ is a result of this strategy and constitutes the centerpiece of the new energy policy. The mandate and autonomy of TÉQ are reminiscent of the status that was attributed to the AEE in 2006. This time, however, the entity is allocated a larger budget and granted greater capacity to leverage new initiatives. The mandate of TÉQ is summarized as follows:

- Coordinate the implementation of all EE, substitution, and innovation programs;

- Offer financing services to energy consumers and businesses, as well as information services to the population;

- Proactively advise the government on standards and regulations, ecofiscal measures, as well as other factors that impact the energy consumption of households, businesses, and government entities;

- Ensure government EE targets are attained and updated as needed;

- Work to reduce the GHG emissions generated by public infrastructure in the education and health sectors by collaborating with stakeholders;

- Ensure accountability by compiling, validating, and disseminating data and information on progress toward achieving targets;

- Identify and address the regulatory and normative barriers to private and public initiatives to attain the objectives of the new energy policy.

While at the time of writing it is too soon to assess the potential impacts this new entity has on the Quebec EE market, its modus operandi will certainly be scrutinized by EE stakeholders across the province and Canada.

Régie de l'énergie

The Régie de l'énergie du Québec[10] is the current public economic regulatory body whose mandate consists mostly of ascertaining whether energy consumers are adequately supplied and charged fair and reasonable rates. More specifically, the Régie reviews and sets annual electricity and natural gas rates based on information gathered during public consultations with energy distributors and consumer associations.

In accordance with law (Act Respecting the Régie de l'Énergie), the Régie must also review all EE programs offered by energy distributors to ensure the programs do not negatively impact energy rates. It evaluates distributor EE programs and strategies and can recommend setting higher or lower targets, as needed. However, the Régie may only observe progress in attaining EE targets and may not demand that new EE measures be implemented.[11]

The Régie also sets the annual duty that distributors must pay to the MERN. These duties were introduced through the Regulation Respecting the Annual Share Payable to the Minister of Natural Resources and Wildlife of 2008 to ensure that government EE programs are funded. The amounts paid for the fiscal year 2016-2017 totaled $44,664,000, with a contribution of $35,941,000 from electricity distributors (Hydro-Québec), $5,677,000 from natural gas providers (Énergir and Gazifère), and $3,046,000 from fossil fuel distributors.[12] These amounts now wholly or partially fund five programs that the BEIE administers, namely Novoclimat 2.0, Rénoclimat, Éconologis, Technoclimat, and ÉcoPerformance. This money also serves to fund other regulatory activities of the Régie.

Energy Distributors

Three main energy distributors operate in Quebec: Hydro-Québec, the state-owned and sole electricity distributor of the province; Énergir,[ii] the main natural gas distributor; and Gazifère, the natural gas distributor that provides services to the City of Gatineau. During the

---

[ii] Previously known as Gaz-Métro (Gaz Métro changed its name to Énergir in November 2017).

last decade, most EE programs have in fact been developed and implemented by these three energy distributors and the MERN (see Figure 12-2). They thus contributed the most to energy reduction efforts.

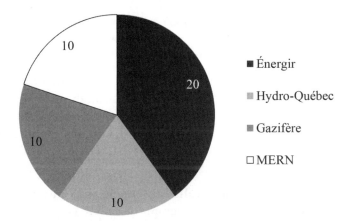

Figure 12-2: Numbers of Main Energy Efficiency Programs Offered in Quebec

Hydro-Québec

Thanks to its leading position on the Quebec electricity market and due to its shared history with the social, industrial and economic development of the province, Hydro-Québec has played and continues to play a key role in the EE market. It remains the biggest investor in EE and, as such, is at the forefront of the province's efforts to reduce electricity consumption. Hydro-Québec considers EE as the ideal way to balance supply and demand despite the EE targets and constraints placed upon it by government policies. As a result, it prioritizes EE over electricity generation or purchases. In fact, Hydro-Québec views EE as a perfect way to curb growth needs and reduce power demand during peak periods, especially since sufficient energy is already available on the grid, but escalations in peak demands already occur and this frequency is expected to increase in upcoming years.[13] For these reasons, the utility decided, as of 2016, to focus future actions essentially on the following three areas: energy savings, demand response, and promoting the use of energy sources most suited to the specific needs of autonomous grids. The goal of Hydro-Québec for 2020 is hence to reduce electricity sales growth by one third through EE initiatives.

Hydro-Québec interventions in EE are categorized into four types of approaches:[14]

1. **Commercial programs** aimed at encouraging clients to reduce energy use or shift energy consumption to off-peak hours. This includes offering incentives and rebates for the purchase of energy efficient showerheads or ENERGY STAR® certified windows through the residential ENERGY WISE program. For the commercial and industrial sectors, grant programs are provided for the implementation of EE or energy conservation projects.

2. **Electricity rates or optional rates** aimed at encouraging clients to decrease or shift energy consumption through price signals. Although no time-of-use tariffs exist in Quebec, special curtailment rates such as the Interruptible Electricity Options allow customers to exercise load-shaving at Hydro-Québec's request during winter in exchange for credits.

3. **Third-party financing or calls for tenders with intermediary organizations** to implement programs or activities. For instance, the BEIE partially funds EE activities and programs.

4. **Other activities** including research and development activities, technology innovation and support for the development of new regulations and standards.

The various strategies formulated in the 2014-2023 Supply Plan (Plan d'approvisionnement 2014-2023) now constitute the foundation upon which Hydro-Québec bases its EE approach to achieve sustainable results and continue the transformation of the EE market. These strategies are summarized as follows:[15]

*   Optimization of existing programs in all business segments through an integrated approach per market;

*   Promotional and innovative initiatives based on an EE awareness and behavioral change approach;

*   Implementation of new demand response interventions;

*   Improvements to energy management offerings for large industrial clients based on a continual improvement process;

*   Optimization of offerings to low-income households;

*   Adoption of an integrated approach specific to each autonomous grid;

- Recourse to external service providers to provide program services which favor a turnkey approach rather than a per-product approach to implementing measures.

This new set of strategies inaugurates a new era for Hydro-Québec given the achievement of the 12-year-long Global Energy Efficiency Plan (Plan global en efficacité énergétique). This plan, launched in 2003, allowed the utility to generate total savings of 8.7 TWh with expenses totalizing $1.7 billion between 2003 and 2015. By surpassing the initial target of 8 TWh by more than 8%, results clearly attest to the robustness of the utility's influence on the market and certainly constitute a solid foundation upon which to build for the future.

Énergir

As the largest natural gas provider in the province, Énergir also plays a leading role in the Quebec EE sector. In the early 2000s, the utility was the first to develop its own Global Energy Efficiency Plan with the aim of encouraging clients to adopt more efficient consumption habits and reduce energy bills. Ever since, Énergir has multiplied efforts to reduce energy waste by proposing several EE programs for every sector. As a result, more than 111,000 EE projects were implemented between 2001 and 2015, representing more than $15 million in subsidies and generating savings equivalent to 420 million cubic meters of natural gas. [16]

Today, with the new 2016 energy policy of the province, EE remains at the heart of Énergir priorities. The utility's long-term vision breaks down into five main objectives:[17]

1. Integrate EE in every activity of the utility;

2. Aim for long-term market transformation;

3. Aim for durability of EE interventions in partnership with other stakeholders;

4. Clearly specify mid and long-term EE targets;

5. Favor efficient and complementary collaborations with other EE stakeholders.

To implement this long-term vision, Énergir mainly elaborates and provides a portfolio of EE initiatives mostly composed of incentive and rebate programs for the purchase and installation of EE technology equipment. As a result, most of the money invested (approximately 85%) in these programs returns to participants in the form of grants, while only a fraction (nearly 15%) serves to pay for administrative expenses. Among the twenty programs currently offered (Table 12-2), five are intangible programs in that they mainly focus on education and awareness activities which generate savings that cannot be precisely measured.

Table 12-2: Portfolio of Énergir EE Programs (2018-2020)[18]

Type of Program	Residential	Commercial, Institutional & Industrial (CII)	Major Industries
Intangible	• Residential awareness • Low-income household supplement	• Business awareness • Low-income household supplement - CII	• Major industry awareness
Tangible	• Electronic and smart programmable thermostats • Hot water boilers • Combined systems • Water heaters	• Mid-efficiency boilers • Feasibility studies • Implementation incentives for EE measures • Condensing boilers • Condensing water heaters • Infrared sensors • Energy innovation • Variable speed hoods • Condensing unit heaters • EE renovations • Solar air preheating systems • New efficient constructions	• Feasibility studies • Implementation incentives for EE measures (Industrial) • Implementation incentives for EE measures (Institutional)

Other Stakeholders

Several actors participate in provincial efforts to promote EE. Various ministries and government bodies have made contributions in their respective jurisdictions to achieve greater EE under the government's sustainable development approach. The ministries in charge of transportation, sustainable development, municipal affairs, and the Régie du bâtiment actively contribute through various initiatives and regulations to the efforts supporting EE and innovation.[19] Other government bodies also work to achieve energy savings by enhancing EE in the infrastructure for which they are responsible.

Various stakeholders in the municipal, private and non-government sectors play key roles in the EE field. For instance, many engineering firms act as ESCOs to offer technical and financial services to implement EE projects under the now widely adopted EPC concept. The use of this unique approach in Quebec has been largely promoted and facilitated by energy consulting firms that work with clients to tackle some of the numerous barriers to EE project implementation. To date, the MUSH sector (Municipalities, Universities, Schools and Hospitals) has been the main beneficiary of these services, while much potential remains in the commercial, residential, and transportation sectors.

POLICIES AND STRATEGIES

2030 Energy Policy
EE has been at the heart of many government initiatives to reduce GHG emissions and transition to a decarbonized economy since climate change concerns have grown over the last decade. The aforementioned action plans, energy strategies, and policies have been aligned in Quebec since 1992 and have helped establish an overall vision for planning targets in energy related matters.

The MERN is in charge of elaborating the energy vision and implementing the new energy policy of the province. Launched in 2016, the 2030 Energy Policy, Energy in Quebec: A Source of Growth, sets ambitious targets including the enhancement of EE by 15% over the next 15 years.[20] Through this policy, the province aims to become a North American leader in the field of renewable energy and EE and build a new, strong and low-carbon economy. The government has therefore planned to assist households, businesses, and public establishments in reducing energy consumption and encourage energy substitution measures through $4 billion in investments over the 15-year period.

The first Action Plan arising from the 2030 Energy Policy and released in 2017 sets concrete actions intended to be put forward by the government to achieve its targets. These actions are grouped into four key directions:
1.   Ensure integrated governance of the energy transition;
2.   Promote the transition to a low-carbon economy;
3.   Offer consumers a renewed and diversified energy supply;
4.   Define a new approach to fossil energies.

The second direction directly impacts the EE sector with specific actions aimed at the ways Quebecers commute, live, innovate and work. The themes and objectives of this key direction are presented in Table 12-3.

Table 12-3: Key Direction 2—Promote the Transition to a Low-Carbon Economy—2030 Energy Policy Action Plan[21]

Theme	Objective
Innovate Green: Support energy innovation and reduced GHG emissions in Quebec	Increase the technological innovation activities of EE, renewable and bioenergy companies.
Move Green: Influence the movement of people and the transportation of goods	Increase the number of electric vehicles.
	Speed up the electrification of public transit services.
	Increase the number of heavy vehicles converted to clean fuel.
	Reduce the petroleum fuel consumption of governmental and para-governmental light vehicle fleets.
	Expand the total offer of replacement fuels: biofuels, natural gas, liquefied natural gas (LNG), compressed natural gas (CNG), propane, hydrogen or electricity.
	Increase recourse to clean energy in businesses, institutions and municipalities.
Live Green: Influence the energy consumption of households and communities	Reduce household energy consumption.
Work Green: Influence the energy consumption choices of business and government	Increase recourse to clean energy in businesses, institutions and municipalities.
	Reduce the energy consumption of businesses, public buildings and municipalities.

GHG Emission Reduction Targets

In the 2013-2020 Climate Action Plan, Quebec targets reducing GHG emissions by 20% below 1990 levels by 2020.[22] This target was, however, revised upward pursuant to the 21st Conference of the Parties (COP21) of the UN Framework Convention on Climate Change that took place in December 2015 (i.e. Paris Agreement). Quebec now pledges to reduce GHG emissions by 37.5% by 2030. This new target is in line with a commitment made in August 2015, a few months prior to COP21, by the Government of Quebec and ten other provinces and states gathered during the Conference of New England Governors and Eastern Canadian Premiers (NEG/ECP). Leaders there pledged to attain a regional GHG emission reduction target of 35 to 45% below 1990 levels by 2030. [23]

Since adhering to the Subnational Global Climate Leadership Memorandum of Understanding in July 2015, Quebec aims to reduce GHG emissions by 80 to 95% below 1990 levels by 2050.

According to the data provided in the 2030 Energy Policy, more than 70% of Quebec's GHG emissions result from the production, transport and consumption of energy. The targets set by the government in this new Energy Policy, including EE targets, should allow a reduction of 16 metric tons in GHG emissions by 2030, representing 18% of 1990 emissions. This is in addition to the 8.5% emission reductions already attained through past and current EE initiatives and other emission reductions generated from non-energy sources.

Table 12-4 summarizes the different GHG emission reduction targets set by the Quebec government and how they compare to the emission reductions that would result from the achievement of the 2030 Energy Policy targets.

Table 12-4: Quebec GHG Emission Reduction Targets

Horizon	Target
Total GHG Emission Reduction Targets Relative to 1990 Levels	
2020	20%
2030	37.5%
2050	80-95%
GHG Emission Reduction Resulting from the 2030 Energy Policy Relative to 1990 Levels	
2030	18%

By adopting the most ambitious targets in Canada, Quebec demonstrates leadership and engagement in the fight against climate change despite its already relatively low carbon footprint. These targets have set a very high bar and, despite the willingness of the government to achieve the 2030 Energy Policy goals, it appears that much effort remains to be deployed in concert with all stakeholders to render these objectives not only realistic but also attainable.

So far, Quebec has achieved GHG emissions about 8% below 1990 levels while provincial GNP has increased by almost 60% since 1990 (see Figure 12-3). This trend is encouraging and demonstrates that economic development and GHG emission reductions can be decoupled largely thanks to EE measures. However despite these preliminary observa-

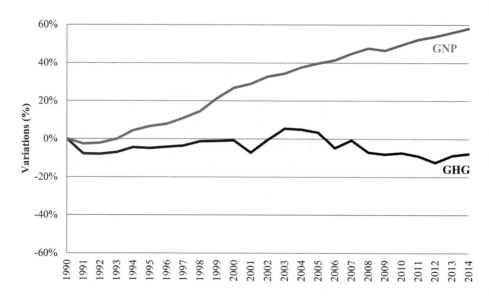

Figure 12-3: Quebec GHG Emissions and GNP Variations Relative to 1990 Levels. Source: MDDELCC and Statistics Canada[24], [25]

tions, much work remains before Quebec achieves its ambitious 2030 and 2050 targets. While the new 2016 energy policy proposes a wide range of initiatives to set the path toward these goals, EE remains if not the most important tool, then certainly one of the most important to achieving government targets.

LEGAL AND REGULATORY FRAMEWORK

Public EE initiatives and, to a larger extent, the energy sector are regulated by several laws codifying and defining the different roles and obligations of the main provincial energy actors. These laws, some of which have already been mentioned, are listed below:

- Act Respecting Energy Efficiency and Energy Conservation Standards for Certain Electrical or Hydrocarbon-Fuelled Appliances (Chapter N-1.01);

- Act Respecting the Régie de l'Énergie (Chapter R-6.01);

- Act Respecting Transition Énergétique Québec (Chapter T-11.02);

- Regulation Respecting a Cap-and-Trade System for Greenhouse Gas Emission Allowances (Chapter Q-2, r. 46.1).

The Act Respecting Energy Efficiency and Energy Conservation Standards for Certain Electrical or Hydrocarbon-Fuelled Appliances was enacted in 2011 and enables the Province, through the Minister of Natural Resources and Wildlife, to set EE and energy conservation standards for any electrical or hydrocarbon-fuelled appliances. To do so, the act bestows upon the Minister the authority to either make mandatory any EE or labeling standards set by a certifying or standards body, or require that appliances be approved or certified by such a body. As such, apart from the building code, this act is the only piece of provincial legislation that directly involves EE equipment and performance.

The Act Respecting the Régie de l'Énergie defines the mandate of the Régie. It was enacted in 1996 and has since been amended many times to comply with the new tasks and duties falling upon the Régie under the different energy policies put into force throughout the years. With the creation of TÉQ, for instance, an amendment to this Act provides that the Régie review the Energy Transition, Innovation and Efficiency Master Plan prepared by TÉQ. It is thus the duty of the Régie to approve the different programs and measures initiated by energy utilities as well as the financial investments necessary for carrying out said initiatives.

The Act Respecting Transition Énergétique Québec was enacted in 2016. It serves as the foundation for the creation of TÉQ and defines its role, mission, as well as its internal structure and sources of funding. Hence, the law provides that TÉQ activities be financed from four main lines of income:

1. The annual contribution received from energy distributors;

2. The sums from the Green Fund provided for under an agreement with the Minister of Sustainable Development, Environment and Parks;

3. The sums from the Energy Transition Fund; and

4. Other sources.

Consequently, TÉQ must file annual financial statements and activity reports for the preceding fiscal year with the Minister of Natural Resources and Wildlife. Activity reports must include the implementation status of the Master Plan, achievement of targets prescribed by the Government, number of programs and measures implemented, and the funds used. These prescriptions allow the government to ensure that TÉQ is totally transparent in its use of public funds and how it manages activities.

The last major regulation, which has indirect implications for the EE sector, is the Regulation Respecting a Cap-and-Trade System for GHG Emission Allowances. Adopted in 2011, this Regulation sets the rules for operating the cap-and-trade system for GHG emission allowances (see rubric opposite). It therefore serves to determine which emitters are required to pay for their emissions, the terms and conditions for registering with the system, emission allowances, and the information that must be provided by registered emitters.

FINANCIAL MECHANISMS

Similar to most provinces and jurisdictions, grants are the main form of incentive offered to tackle the financial barriers to EE project implementation in Quebec. Indeed, most funding provided by the Province is based on non-repayable grants from the government or energy distributors. While some repayable funding mechanisms such as low-interest loans or financial leases exist to finance EE projects, these mechanisms have often been considered inadequate or ineffective to meet the needs of market actors.

As a result, the Quebec government, through the AEE and the BEIE, granted approximately $500 million, as listed in Figure 12-4, between 1997 and 2014 through EE programs. In contrast, for the 2008 to 2014 period, Quebec energy utilities invested slightly above $1 billion in EE, with Hydro-Québec contributing about 92% of that sum (see Table 12-5).[29] Most of that investment went to financing the DSM programs put in place by the major organizations on the market.

Dividing the values presented above by the energy savings achieved during the same period results in an average cost of $122 per MWh of energy saved. Compared to the levelized cost of different electricity sources in Canada (see Figure 12-5), this clearly demonstrates that EE is one of the cheapest sources of energy, in particular when compared to other types of renewable energy such as solar photovoltaic, geothermal and offshore wind energy. This average cost is comparable

Quebec Cap-and-Trade System for Emission Allowances[26]

A cap-and-trade system (CTS), also known as a carbon market, is a market-based approach to controlling GHG emissions by inducing a carbon cost in business decision-making. This cost for "having the right to pollute" is intended to encourage the implementation of clean technologies as well as EE measures to decrease GHG emissions.

On 1 January 2013, Quebec launched its own cap-and-trade system under the Western Climate Initiative (WCI) carbon market and associated the system with California's one year later. This marked a new era for the province in the fight against climate change. The system primarily targets businesses that annually emit 25,000 metric tons or more of CO_2 equivalent, but individuals or entities that would like to participate in the carbon market are also allowed. Solely the industrial and electricity sectors were subject to the system for the first compliance period (2013-2014), but since 2015 fossil fuel distributors have also been subject to the carbon market. To comply, emitters must purchase allowances corresponding to the emissions they intend to produce. These allowances can be purchased from other emitters, from the government through auctions, or traded with other participating entities if the total emissions of registered emitters are below allowances. All auction proceeds go to the Quebec Green Fund and are earmarked to finance the initiatives contained in the 2013-2020 Climate Change Action Plan.

To increase the price of emission allowances and spur GHG mitigation efforts, the Government establishes a yearly cap on total emission units put into circulation. As a result, the minimum price per emission unit, set at $10.75 per metric ton in 2013, is scheduled to increase at a rate of 5% plus inflation every year until 2020. Given the minimal impact this system has had on total GHG emissions, many argue that the carbon price is much too low for decision makers to factor this into corporate policy. With Ontario joining the CTS in 2017, it will be interesting to monitor how this affects the market.

to certain fossil fuel and hydropower plants. Contrasting this value with the average cost of running efficiency programs in the United States (about $30/MWh[31]) demonstrates that there is much room for improving how funds are used in Quebec to leverage EE initiatives and generate energy savings. Despite this, investments made in the EE sector have had an overall positive impact on the market, as illustrated in Figure 12-6, resulting in net energy savings of about 1.7 million tons of oil equivalent (Mtoe) between 1990 and 2014.

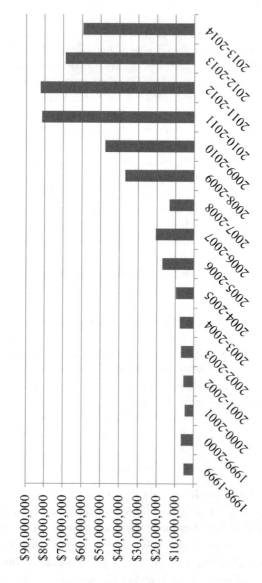

Figure 12-4: Financial Aid Granted by the BEIE (AEE before 2011). Source: AEE financial statements[27] and rate cases submitted to the Régie[28]

Table 12-5: Quebec Energy Utility Investments in Energy Efficiency and Innovation—2008 to 2014 ($)[30]

	Hydro-Québec	Énergir	EEF[iii]	Gazifère
2008-2009	155,700,000	8,500,000	4,900,000	300,000
2009-2010	177,300,000	10,400,000	2,600,000	400,000
2010-2011	178,200,000	10,700,000	3,300,000	400,000
2011-2012	179,100,000	10,400,000	3,100,000	400,000
2012-2013	150,000,000	11,100,000	-	300,000
2013-2014	127,000,000	15,000,000	-	500,000
Total	967,300,000	66,100,000	13,900,000	2,300,000

[iii] Énergir's Energy Efficiency Fund.

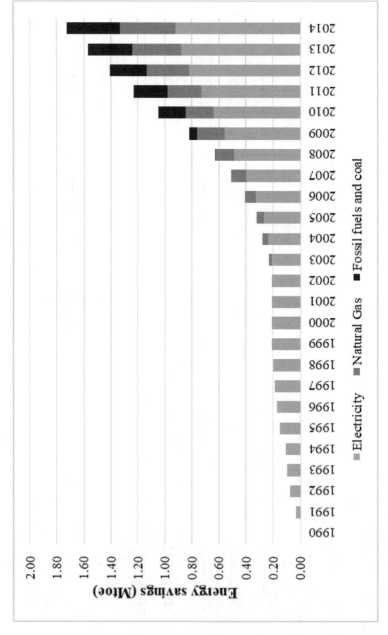

Figure 12-5: Levelized Cost of Electricity[32]

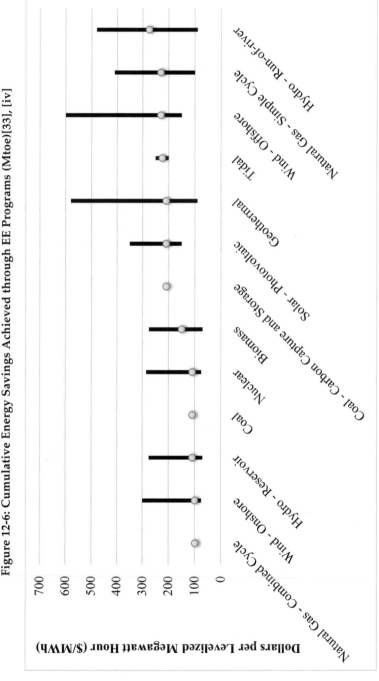

Figure 12-6: Cumulative Energy Savings Achieved through EE Programs (Mtoe)[33], [iv]

[iv] Million tons of oil equivalent (1 toe = 41.868 GJ).

With the creation of TÉQ and a budget allocation of $4 billion for the next fifteen years, uncertainty remains as to how these funds will be used to foster EE initiatives. Nonetheless, given the market's reliance to non-repayable financing mechanisms (grants and rebates), it is very likely that this form of financing will prevail for some time.

Less traditional financing mechanisms such as low interest loans or loan guarantees are still very sparse in Quebec. One such example is the Biomass Energy Fund (Fonds Biomasse Énergie) administered by Fondaction, a labor-sponsored fund that aims to support fossil-fuel-to-biomass conversion projects by providing support as well as financing in the form of loans. Other similar funds are also available for EE project implementation through banking institutions such as the Royal Bank of Canada (RBC) Energy Saver™ Loan,[34] or the Desjardins Energy-Efficiency Loan[35] which offers the possibility of postponing repayment and basing loan terms and conditions on projected savings. These kinds of financing products are, however, still relatively unknown to the general public and thus have very little impact on the EE market.

Chapter 13

Saskatchewan

Ms. Brenda WALLACE, City of Saskatoon

ENERGY SECTOR

During the 1995-2006 period, energy consumption in Saskatchewan decreased by 4%. In the ensuing 10 years, however, annual consumption increased at a very high rate to reach an average of 11,748 ktoe, 33% more than in the 1995-97 period. This drastic increase was mainly due to the energy consumption of the transportation and industrial sectors, which increased by 46 and 54% respectively. In 2013-15, industrial sector annual energy consumption accounted for 29% of total energy consumption, whereas the transportation sector accounted for 37%. The agriculture sector consumed third most during this three-year period with 15% of overall energy consumption. This sector underwent a significant 37% increase compared to 1995-97 levels, and 75% more than 2001-2003 levels. The residential sector accounted for 10% of overall consumption in the province in 2013-15, while the commercial sector consumed 8%. Finally, the public administration sector consumed the least, 2%, during the 2013-2015 period (See Figure 13-1).

ENERGY EFFICIENCY BACKGROUND

The global oil crisis of the late 1970s led to concerns about energy security, resulting in more energy conservation and oil independence efforts. Rob Dumont at the Saskatchewan Research Council developed HOTCAN, an energy modeling software program used for planning and building energy efficient homes in Canada. Robert Besant and Carey Simonson developed a plate air-to-air heat exchanger at the University of Saskatchewan. Dirk vänEE and Rick Olmstead started a company using the idea of an air-to-air heat exchanger to develop a heat recovery ventilator (HRV) that radically transformed ventilation and heat recov-

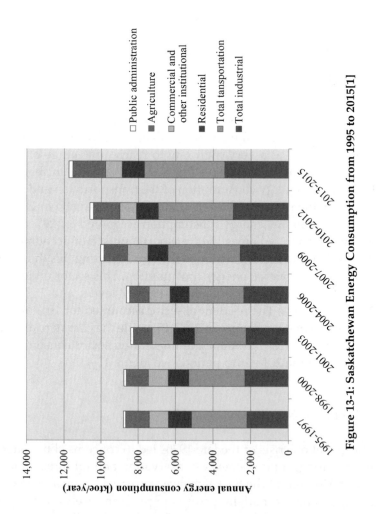

Figure 13-1: Saskatchewan Energy Consumption from 1995 to 2015[1]

ery systems in North America.[2]

The Saskatchewan Conservation House, constructed in 1977 and whose primary project manager was Harold Orr, is believed to be one of the first conservation demonstration houses in North America.[3]

Figure 13-2: Saskatchewan Conservation House, Encyclopedia of Saskatchewan

This project pioneered heat recovery ventilation, air tightness testing, high insulation levels and drain water heat recovery. The Saskatchewan Research Council led the project and partners included the Saskatchewan Housing Corporation, SaskPower, the National Research Council, University of Regina, and University of Saskatchewan. Robert Besant was also heavily involved in the project and has stated that the greatest legacy of the Conservation House was to increase insulation standards in Canada. "The demonstration house influenced thousands of contractors who went on to build R-2000 homes and led to improvements in windows and doors," claimed Besant. Many Canadian innovations and EE programs have been significantly influenced by the development and testing of this home.

In the 1990s, the Saskatchewan Energy Conservation and Development Association (SECDA) published studies and reports on energy alternatives and efficiencies. In July 1994, SECDA published a feasibility study on nuclear power plants as an option for future electrical generation in Saskatchewan.[4] The organization also researched and advocated for provincial energy conservation.

In 2007, the Saskatchewan Research Council built the Factor 9 Home, a single-family residence whose project target was to reduce the energy intensity of the average existing Saskatchewan home (circa 1970) by a factor of 9 and reach 30 kWh/m2.[5] Key performance features included an air tightness target of 0.5 air changes/hour at 50 Pa.[6] Additionally, water efficiency and other advanced environmental features were showcased and the design targeted a factor of 2 in water consumption reduction compared with conventional homes.[7] The project met both its energy and water use performance targets.

In 2011, following in the footsteps of the Saskatchewan Conservation House and the Factor 9 Home, the VerEco Demonstration Home was built and exhibited for one year at the Western Development Museum in Saskatoon. The home was designed to be NetZero energy and demonstrate reduction in construction waste, GHG emissions, and a high-quality indoor environment. The museum exhibit enabled education activities that included public tours, school tours, an outreach program and the Expert Series. Over 5,500 people took part in the program. Public tours provided information on the design and construction of energy efficient homes. The two-hour school tour consisted of an exploration of the VerEco Home, a scavenger hunt in two related exhibits, Winds of Change: Wind Energy and The Sod House from the Winning the Prairie Gamble, as well as on-site information stations on the various green technologies used in the NetZero Home. The outreach program included presentations about energy-efficient home construction and design to groups throughout the province. The Expert Series comprised weekly public seminars during which local and provincial experts shared their EE and home design expertise.

The German PassivHaus program, resulting in the Temperance Street Passive House, was also significantly influenced by the Saskatchewan Conservation House.[8] In April 2015, the Passive House Institute in Germany honored Harold Orr and his team with a pioneer award. Significantly, the Temperance Street Passive House was the first Passive House project certified in Saskatchewan, 39 years after the construction of the Saskatchewan Conservation House. The Temperance Street Passive House uses electric heat solar and photovoltaic panels.[9] The home also uses a 93% efficient heat recovery ventilator, achieved an air tightness level of 0.37 air exchanges per hour during the blower door test, utilizes sealed and taped plywood, and has R-65 above grade walls and an R-105 attic.

STAKEHOLDERS

The two main utilities in Saskatchewan are SaskPower and SaskEnergy, both of which are provincial Crown Corporations.

SaskPower is the main electric utility and services more than 522,000 customers. In 2013, it launched the Industrial Energy Optimization Program designed to assist energy-intensive industries in identifying energy waste in operations to reduce electrical use and demand costs.[10] The intent was to create a win-win strategy that benefited industry and helped balance provincial energy supply and demand.

SaskEnergy distributes natural gas to 390,000 customers. It offers EE programs such as the Commercial Boiler Program, Commercial HVAC Program, and the ENERGY STAR® Loan Program.[11]

Other utilities are operated by the cities of Swift Current and Saskatoon despite the Power Corporations Act of Saskatchewan[12] which legislates that SaskPower is the only entity in Saskatchewan permitted to transmit power from one property to another. Since the two cities already had existing power corporations when the act came into force, they were allowed to maintain their existing franchise areas.

The City of Swift Current provides electrical services to the majority of its residents and purchases power from SaskPower as a reseller. The city has no known EE or renewable energy programs.

The City of Saskatoon operates an electrical utility called Saskatoon Light and Power (SL&P), a division of the city that purchases power from SaskPower at a reseller rate.[13] The franchise area of SL&P mostly includes the areas of Saskatoon constructed prior to 1958.[13] The SL&P rate matches SaskPower's to ensure all citizens in Saskatoon pay the same rate for power. SL&P also offers programs that emulate the SaskPower Small Power Producer and Net Metering programs. SL&P installs LED lighting in new areas of Saskatoon. The inexpensive and quality LED fixtures have been incorporated into the SL&P lighting standard.

SL&P has also begun installing water and electricity smart meters which enable customers to accurately monitor and measure the effects of any EE improvements in their homes or businesses. Over 51,000 electricity smart meters were installed in 2017, while over 14,000 water meters have been upgraded and deployment is scheduled for completion in 2021. SL&P also allows customers to borrow an in-home display that provides real-time data by communicating directly with smart meters. Over 75 customers have taken advantage of the in-home display loaner program to date, including families involved in Student Action for a Sustainable Future projects. The City of Saskatoon Corporate Revenue division is also working on implementing an online interface to allow electricity and water customers to access their detailed smart meter consumption data online. Consumption profiles will include weather data and various comparative features.

In 2017, SL&P commissioned a Solar Power Demonstration Site, a collaboration project between SL&P, the Saskatchewan Environmental Society (SES), Saskatchewan Polytechnic, and the SES Solar Cooperative. The project is located at the City of Saskatoon Landfill Gas Power Generation Facility and consists of four ground-mount arrays with a total of 92 solar panels with a 30.66 kW generation capacity. The project provides real-time production monitoring displayed on the City of Saskatoon website.

SES Solar Cooperative Ltd. is a consumer cooperative established by the Cooperatives Act of 1996. Members of the Cooperative purchase one common share and one or more preferred shares and own a portion of the entire operation rather than one or more specific solar panels.[14] In 2017, the organization was granted the right to allow members to invest in shares in the form of a Registered Retirement Savings Plan (RRSP).

SL&P also recently agreed to partner with Sun Country Highway, the Saskatchewan Environmental Society, Saskatchewan Research Council, Saskatoon CarShare Cooperative, and the University of Saskatchewan Car Share Program to conduct an electric vehicle (EV) car-share pilot project.[15] Sun Country Highway offers a charging network and sells electric vehicle equipment. It operates approximately 25 charging stations in Saskatchewan and hundreds throughout North America.[16] The pilot project utilizes virtual net metering to enable EV drivers to disclose that the energy used to charge the EV comes from solar power.

The Western Development Museum (WDM) in Saskatoon features the Winds of Change exhibit to highlight the past, present, and future of wind turbines in Saskatchewan.[17] This initiative highlights the importance of wind turbines in the province and their role in supporting alternative energy production to meet electricity demand. WDM also features the Fueled by Innovation exhibit which explores the alternative fuels that power vehicles in Saskatchewan.[18]

The university sector has also played a major role in the development of EE in the province. In the late 1970s, the University of Saskatchewan started installing digital building controls to monitor the mechanical and electrical systems in facilities. From 2007 to 2013, the university carried out on-campus lighting retrofits that reduced lighting energy consumption by approximately 23%.[19] Other initiatives include hiring to fill energy manager and commissioning positions, as well as installing controllers for parking lot plug-ins and 24-kilowatt solar panels at the Horticulture Science Field Facility which supplies 70% of the building's electricity load.[20]

At the College of Engineering, University of Saskatchewan, most EE training occurs at the MSc and PhD levels. The college conducts research on energy efficient energy exchangers for building HVAC systems and is a global leader in air-to-air heat recovery ventilator research. Notable researchers associated with the college include Robert Besant and Carey Simonson. Current research focuses on solid and liquid desiccants for cooling and dehumidification. According to the Intergovernmental Panel on Climate Change, the need for cooling is expected to increase by a factor of 30 throughout this century due to climate change and the increased demand for indoor comfort in developing countries.

Current EE initiatives in the health care sector are also of interest. The Saskatoon Health Region and SL&P are conducting a feasibility study of a combined heat and power (CHP) plant at St. Paul's Hospital.[21] The CHP plant would efficiently and simultaneously generate heat and electricity, which would contribute to meeting the hospital's heating and electrical requirements while reducing operating and maintenance costs.

Other EE actions have been carried out in the health care sector. Pursuant to the announcement from the Saskatchewan Office of Energy Conservation to mandate EE in public buildings, the Ministry of Education notified school divisions in 2008 that all new schools had to be designed to target LEED Silver Certification with a minimum energy performance of 30% above the MNECB 1997.[22]

The Saskatchewan Urban Municipalities Association operates other initiatives in the municipal sector. For instance, it offers a solar pool heating program[23] developed between 2003 and 2006 in collaboration with the Saskatchewan Office of Energy Conservation. Under this program, the community of Bengough reduced energy use by 60% and prevented its pool from being closed due to rising operating costs.[24]

POLICIES AND STRATEGIES

According to a statement from SaskPower,[i] the previous Federal Government (2006-2015) decided that regulatory oversight should be by industry rather than a blanket countrywide approach. Coal-fired electricity generation was one area for which regulations were developed and with which SaskPower must comply. Details are outlined on the Environment and Climate Change Canada website. This regulation requires that when a coal-fired turbine reaches 50 years of age, deemed its economic life, it must be either decommissioned or refitted with carbon capture and sequestration (CCS) equipment designed to ensure it emits no more GHGs than a natural gas-fired generator (representing about a 60% reduction). These regulations came into effect on 1 July 2015. The current federal government has adopted these regulations but has established the final year for conventional operations to be 2030 no matter the age of coal-fired plants. This will affect the Shand Power Station which fulfils its economic life in 2042.

SaskPower recognized a long time ago that GHG regulations were very likely to be implemented. Hence, management has been exploring various CCS systems for nearly 25 years. In the early 2000s, a detailed assessment was carried out on an oxyfuel system that would have included a greenhouse at the existing Shand Power Station. The costs were daunting and this approach was abandoned. Subsequently, retrofitting an existing plant with a post-combustion amine-based system was investigated. This approach was eventually adopted, resulting in the Boundary Dam Power Station Unit 3 (BD3) project that was completed in October 2014. It is operating successfully and has a 30-year economic life that is aligned with federal requirements. BD1 and BD2 were decommissioned since they were considered too small to be retrofitted.

SaskPower is now drawing from the lessons learned in the BD3 ex-

---

[i] Personal correspondence with Ian Yeates, Director, Supply Development Carbon Capture, SaskPower.

perience and assessing the potential to retrofit the remaining coal-fired fleet over the next 15 to 20 years. The sites that have to be retrofitted or retired in the near term are BD4 and BD5, which are both similar to BD3.

At this time, provincial authorities are negotiating an equivalency agreement with federal counterparts to allow the Province to oversee SaskPower compliance with federal objectives. The intent is to allow Saskatchewan to treat electrical coal-fired power assets as a fleet and remove the 50-year age limit per asset. This will allow SaskPower to optimize its investments in CCS and minimize increases to the rate base. An agreement should formally be concluded in early to mid-2018.

LEGAL AND REGULATORY FRAMEWORK

Saskatchewan has no specific EE laws that apply to the general public, other than federal EE standards.

With regard to building codes, the Province of Saskatchewan is scheduled to enforce the NECB 2015 and Section 9.36 of the National Building Code on January 1, 2019.[25]

In 2005, the Government of Saskatchewan enacted the Energy-efficient Household Appliances Provincial Sales Tax (PST) Remission and Exemption Regulations.[26] The regulations provide a PST exemption on refrigerators, freezers, dishwashers and clothes washers that are certified as ENERGY STAR® household appliances by the NRCan Office of Energy Efficiency.[27]

As for the implementation of a carbon tax or cap-and-trade system, the Saskatchewan government has clearly expressed its opposition to such programs and, accordingly, some degree of uncertainty exists as to how federal policy will affect SaskPower[i] and energy consumers.

Carbon taxes are straightforwardly applied – corporations pay for GHG emissions on a per ton basis. This therefore increases the cost of doing business, which in turn leads to higher rates to pay for the tax. The purpose of carbon taxes is to make fossil fuel-based electricity generation more expensive and so incent utilities such as SaskPower to reduce reliance on such resources. The economic consequences of prematurely writing off existing serviceable assets and investing in alternative power generation assets are significant. From a consumer and business perspective, the more costly electricity becomes, the greater the incentive for energy efficient equipment and appliances. This involves a slow and expensive transformation for both households and businesses, but per capita electricity consumption presumably falls over the ensuing decade or two. The consequences for Saskatchewan are significant in that a number of critical industries are heavy electricity users, and these might become uncompetitive due to such a change in SaskPower's rate structure.

A cap-and-trade system is a variant on the carbon tax and essentially consists of government regulating the allowable emissions from individual enterprises and households, i.e. the cap. Any entity that can reduce its emissions below the cap would have credits to sell, and any entity that cannot reduce emissions would need to buy credits to comply with the policy, i.e. the trade. This system is hampered by its complexity since it is difficult to define what credits are and the role of government in determining what should be the levels of allowable emissions. Furthermore, temptation to game the system and the opportunity to do so are high. Consequently, these systems have not been successful elsewhere and the provincial government is also opposed to this approach.

FINANCIAL MECHANISMS

A few EE and clean energy financing initiatives have been established in Saskatchewan over the years.

One of the earliest was a collaboration between SaskPower and Honeywell. The two organizations delivered joint EPC projects under a strategic alliance partnership from 1999 to 2015. Projects included the Delta Bessborough Hotel, T&T Towers, City of Regina, and several school divisions. Many EPC projects were completed in schools, healthcare facilities, and in Saskatchewan Property Management Corporation facilities. Typical paybacks ranged from seven to 11 years. SaskPower and Honeywell renewed their partnership in 2010 for five years, but did not renew thereafter. The 2010 press releases announcing the partnership extension stated that over 200 facilities had been retrofitted through the partnership.[28] At that time, the organizations reported annual savings of more than $5.1 million and over 32 million kilowatt hours of electricity.

The Government of Saskatchewan promoted another EE financing initiative, the Go Green Fund which comprised a $60 million commitment between 2008 and 2015 to help people, communities, non-government organizations and businesses address environmental issues.[29] It included grant programs for solar panels, toilet retrofits, and energy efficient new homes. For instance, the Energy Efficiency for New Homes program, administered by SaskEnergy on behalf of the government, provided a $2,400 grant to residents who built or purchased a newly constructed ENERGY STAR® qualified or R-2000 certified home.[30]

Furthermore, SaskPower currently offers various EE programs for the residential, commercial and industrial sectors, namely the Commercial Lighting Incentive, Commercial Refrigeration Incentive, Compressed Air System Program, Demand Response Program, Industrial Energy Optimization Program, Municipal Ice Rink Program, Parking Lot Controller Program, SaskPower Efficiency Partners, Solar or Wind-Powered Water Pump Grant, Walk-Through Assessment Program, as well as the Commercial Boiler Program and Commercial HVAC Program.[31] [32]

Other than traditional EE grant programs, SaskEnergy and Sask-Power jointly offered an ENERGY STAR® Loan Program.[33] The program ran from July 1, 2016 to June 30, 2017 and offered customers financing between 0% to prime +2% on the purchase and installation of eligible ENERGY STAR® certified equipment. Eligible equipment included furnaces, boilers, heat recovery ventilators, air conditioners, and domestic hot water heaters.

As for private initiatives, Affinity Credit Union (Affinity) offers EE retrofit loans to non-profit organizations.[34] Retrofit loans cover expenses such as energy audits carried out by qualified engineers, materials and labor needed to complete retrofits, and project management or technical assistance expenses. Affinity offers the program in partnership with the SES. The SES performs the energy audit, while Affinity also offers a specialized lending program for community projects and organizations including non-profit social enterprises and cooperatives. This program covers the renovation of community-owned facilities, as well as projects that enhance environmental sustainability and promote ecologically responsible practices.[35]

Unfortunately, other emerging EE and clean energy financing mechanisms such as Property-Assessed Clean Energy (PACE) and on-bill utility financing programs are prohibited by the Cities Act[36] and Local Improvements Act[37] of Saskatchewan.

Chapter 14

Yukon

Ms. Stephanie WHITEHEAD, Yukon Energy Corporation

ENERGY EFFICIENCY BACKGROUND

The extreme climate of the Yukon results in high energy use and costs, yet many energy uses such as heating are not voluntary. Thus, managing energy use can have a significant impact on household, business and institutional budgets.

The territorial and municipal governments as well as the electrical utilities offer Yukoners options to help use energy more efficiently through rebate programs, financing opportunities, building standards and education. The remoteness and cold climate of the territory make the adoption of many popular EE products costly or unfeasible, but many of these challenges are outweighed by the significant cost savings that can be realized.

Programs to date have mainly focused on the residential sector, with some programs recently introduced for the commercial and institutional sector.

STAKEHOLDERS

The Yukon Government Energy Strategy, released in 2009, informs the goals of the territorial government on a number of energy issues, one of the priorities being efficiency and conservation.

With direction established in the Energy Strategy, the Yukon Government offers a suite of incentive programs to encourage EE in the residential and commercial sectors under the Good Energy program. The Energy Branch Energy Solutions Centre acts as a storefront for Yukoners looking to access these programs and learn about energy and EE in general.

Yukon Housing has worked to develop capacity in the territory's construction industry to build energy efficient homes, providing energy efficient public housing and offering a low-interest home repair loan program which includes EE upgrades.

Also, the City of Whitehorse introduced EE standards in its building bylaws, while reduced energy use and GHG emissions are goals established in the Whitehorse Sustainability Plan.

There are two electrical utilities in Yukon, Yukon Energy Corporation and ATCO Electric Yukon. Yukon Energy is the primary generator and ATCO Electric is the primary distributor. The two utilities administer inCharge, an electricity conservation program launched in 2014. Both utilities are regulated by the Yukon Utilities Board.

POLICIES AND STRATEGIES

The Yukon Government Energy Strategy recognizes efficiency and conservation as key components to a sustainable and dependable energy sector. The Energy Strategy priorities for efficiency and conservation are:

- Increase EE in Yukon by 20% by 2020;

- Reduce energy consumption in Yukon buildings;

- Reduce energy consumption in the transportation sector;

- Promote the use of energy efficient products by providing incentives for products that meet energy performance standards; and

- Improve EE in Yukon government operations.

The Yukon government joined the Pan-Canadian Framework on Clean Growth and Climate Change in late 2016. Yukon's annex includes a commitment to partnering on investments in EE improvements to buildings, as well as renewable energy, research and pilot projects.

Yukon Housing offers a breakfast seminar series on building science that includes EE topics. Energy Solutions Centre also offers training on a variety of energy related topics, some of which address EE.

There are some residential and commercial LEED certified buildings in Yukon and a number of local design firms are LEED accredited professionals.

LEGAL AND REGULATORY FRAMEWORK

The City of Whitehorse introduced EE into its Building and Plumbing Bylaw in 2012 to ensure energy is used efficiently and sustainably so that homeowners could lower energy bills. These EE requirements are among the most stringent in the country. The EE requirements in the city bylaw are periodically reviewed and updated. The latest update introduced a requirement for new homes to have an EnergGuide rating label. The bylaw also required that design for heat recovery ventilators be conducted by a designer certified by the Heating, Refrigeration and Air Conditioning Institute of Canada.

In 2015 the Yukon government adopted Section 9.36 of the National Building Code with provisions for log home construction, unheated buildings and minimum ceiling insulation upon the recommendation of a special advisory committee to ensure that the standards are relevant and appropriate to the types of construction in Yukon. This did not affect the City of Whitehorse bylaw.

Electrical utilities are regulated by the Yukon Utilities Board which approves DSM programs and budgets thereof. The utilities evaluate their DSM programs and submit a report to the Yukon Utilities Board which, in turn, has it reviewed by a third-party evaluator.

FINANCIAL MECHANISMS

Yukon Government Energy Solutions Centre

The Yukon Government Energy Solutions Centre offers programs to support EE in both the residential and commercial sectors.[1] These include the Good Energy Program for the residential sector, the Commercial Energy Incentive Program, as well as the StartPoint Energy Audits for commercial and institutional customers.

The Good Energy Program offers residential customers home energy kits, a fridge retirement incentive, ENERGY STAR® appliance incentive, home energy assessments, super-insulated new home incentive and a suite of energy upgrade incentives for existing homes.

Incentives of $100 to $250 are offered for home energy assessments completed by NRCan certified energy advisors. The contact information of energy advisors is provided on the Good Energy program website.

Another program includes the Quick Start Home Energy Kits

program. Easy-to-use products to conserve water, electricity and heat are provided for free and include window film, a CFL bulb, low-flow showerhead, bathroom faucet aerator, fridge/freezer thermometer, kitchen faucet tap aerator, shower timer, hot water gauge, LED night light, plumbing Teflon tape, and weather stripping. A detailed installation guide and information sheet are also provided with the kit.

The Refrigerator Retirement Program offers homeowners $50 to retire currently-used refrigerators that are older than ten years. The program includes free pick-up and transportation to the landfill and pays for the landfill tipping fees.

The Good Energy Program offers up to $100 per unit as an incentive on ENERGY STAR® refrigerators, freezers, clothes washers and ventless clothes dryers to participants in communities connected to the hydro grid. It provides up to $300 to participants in communities not connected to the hydro grid. Incentives representing 20% of the cost up to $1,200 are also offered on Canadian Standard Association (CSA) approved solar domestic hot water heating systems and $150 on drain water heat recover systems. Incentives are also offered on efficient home heating systems fired by either wood, oil, propane or electricity. Eligibility requirements and incentive rates vary depending on the type of heating system.

For homeowners renovating homes, the Energy Renovations program offers incentives on heat recovery ventilators ($500), ENERGY STAR® windows and doors ($100 per unit up to $1,000), and insulation (incentives based on area insulated and insulation levels).

An incentive of up to $10,000 is offered for new homes that meet at least a 50% improvement over a National Building Code 2015, which is approximately equivalent to the now discontinued EnerGuide 85 rating. Energy performance modeling and application submissions are completed with staff at the Energy Solutions Centre and an NRCan certified energy advisor.

The Commercial Energy Incentive Program offers incentives for LED lighting upgrades in commercial or institutional buildings and thermal enclosure upgrades for multi-family residential buildings.

The LED lighting upgrades for Commercial Buildings Program offers a 20% rebate on LED lamps, fixtures and controls and a 40% rebate on exterior LED lamps, fixtures and controls up to $10,000. Lamps and fixtures must be ENERGY STAR® or Design Lights Consortium qualified and exterior lights must meet Dark Sky criteria.

The Thermal Enclosure Upgrade for Mixed Use Buildings program provides an incentive of up to $50,000 for energy assessments, thermal enclosure upgrades, heat recovery ventilator installations, as well as ENERGY STAR® windows and doors. Incentive levels depend on the upgrades installed.

The Energy Solutions Centre also offers the Start Point Energy Audit programs for commercial and institutional buildings to help participants understand their energy consumption and expenses. The audit also serves to advise on the initial course of action to address any issues and increase building EE. The audit includes analyzing building energy demand and consumption while taking into account weather, occupancy, operating schedules, climate zones, as well as building type and construction. Audit results are then compared with the typical energy use of similarly sized Canadian buildings.

inCharge

The territory's two electrical utilities, Yukon Energy Corporation and ATCO Electric Yukon offer the inCharge electricity conservation program to residential electrical customers. The utilities were directed by the Yukon Utilities Board to prepare a DSM program in consultation with stakeholders. A conservation potential review (CPR) was completed in partnership with the Yukon Government in 2011. Based on the findings of the CPR, a suite of programs for the residential, commercial and institutional sectors was developed. In addition, a Communications and Engagement Program, Evaluation Measurement and Verification Plan, and Program Implementation Plan were proposed. The programs were presented to the Yukon Utilities Board, and two residential programs were approved and then launched in late 2014.

The first residential program offers rebates on ENERGY STAR® LED bulbs and mechanical block heater timers, distributes electricity saving kits, as well as provides education and engagement for customers on electricity conservation in general.[2]

The second residential program offers rebates of up to $7 on ENERGY STAR® LED bulbs and up to $10 on mechanical block heater timers purchased from any Yukon retailer. Rebate forms are submitted either at participating retail stores, on the inCharge website, by email, snail mail, or at utility offices. Customers receive their rebate as a discount on their electrical bill.

Energy saving kits are distributed for free at community events or to customers that request assistance in reducing electricity bills. When delivering the kits to customers, inCharge staff use the opportunity to engage with the recipient on general electricity conservation and inform them about the inCharge program. The kits contain LED bulbs, a smart power bar and a block heater timer. Single LED bulbs are also distributed under the program.

The inCharge website was designed to provide information on offered programs as well as electricity conservation in general. The website was redesigned in 2016 and now includes an option to submit rebate applications and offers additional energy conservation tips. Point of purchase material, rebate forms and event banners were also redesigned, and an online advertizing campaign was developed to promote the inCharge website and increase brand awareness.

Yukon Housing offers a Home Repair Loan program[3] intended to improve the quality and sustain the useful life of the Yukon housing stock. Eligible homeowners can borrow up to $50,000 to address health and safety concerns, accessibility needs and/or improve home EE. Loans are offered at the prime rate plus 1% amortized over fifteen years.

Part II

Best Practices and Case Studies

Chapter 15

Energy Performance Contracts (EPCs)

Mr. Peter LOVE, Energy Services Association of Canada

DESCRIPTION

Energy performance contracts (EPCs) have been successfully used in Canada for over 35 years to increase EE in existing buildings, particularly publicly owned municipal and other government buildings, as well as education facilities and hospitals.

EPCs are agreements, between an end user and an energy service company (ESCO), which guarantee energy savings arising from EE upgrade projects. These savings serve to finance initial capital costs over the course of projects. EPCs thereby capture future energy savings from energy retrofits to finance global project cost. Most importantly, they transfer the financial and technical risks associated with major EE upgrades to third parties.

Figure 15-1 illustrates how an EPC works. Prior to a contract, some of the energy used by a given building is wasted on inefficient uses. During the contract, reduced energy costs are used to finance up-front capital costs. By the end of the contract, the end user pays a lower energy bill during the effective useful life of the new equipment.

Specific benefits of using EPCs include the following:

- Turnkey—One contract covers a wide range of products and services;

- Comprehensive—Achieves much deeper energy savings than traditional approaches;

- No up-front capital required (in some models);

- 35 years—EPCs have been used for a long time;

- Immediate recognition and resolution of problems;

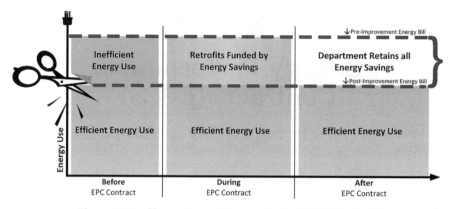

Figure 15-1: Illustration of How a Typical EPC Works[1]

- Financed by energy waste, so taxpayers never make up any losses;

- Cost effective—Based on 25 years of experience with these contracts, NRCan has concluded that they are NOT more expensive than traditional mechanisms;

- Includes measurement and verification of savings.

Further information on EPCs and their applications around the world is available in World ESCO Outlook.[2]

HISTORY

Econoler was founded by Hydro-Québec and a Quebec-based engineering firm in 1981 as the first ESCO in Canada and one of the first in the world. The firm developed a new concept based on a shared savings approach whereby a contract signed under an open-book agreement could be terminated upon complete payment of all project costs (fast-out approach) even when paid in full prior to the end of contract term.

The Econoler ESCO market rapidly developed over a period of ten years under this new concept, thanks to a partnership with Petro Canada that enabled the creation of subsidiaries throughout the country.

The market slowed down in the early 1990s when utilities began massively implementing DSM programs in many Canadian provinces. ESCOs remained active nevertheless, mainly thanks to the leadership of the federal government through the Natural Resources Canada Federal

Buildings Initiative (FBI) developed in 1991 as an effort to harness and capitalize on existing EE potential in federal facilities.[3]

In the 2000s, the EPC mechanism was used again to increase EE in existing buildings, particularly publicly owned buildings at municipal, provincial and federal levels.

MARKET

Since most companies that offer EPCs in Canada also provide other services, it is difficult to estimate the current size of the Canadian EPC industry. The most recent compilation by the Energy Services Association of Canada states that 280 EPC projects have been completed over the last 10 years.[4] Projects range in size from $1 million to $50 million and revenues are estimated at nearly $300 million for projects carried out under a performance guarantee.

Table 15-1 presents the distribution of the 280 projects by sector.

Since all the projects in the apartment/condo sector were for public housing agencies, more than 90% of the projects undertaken in the last 10 years have been carried out for public institutions and most

Table 15-1: EPC Projects in Canada by Sector: 2006-2016
Source: Energy Services Association of Canada Guaranteed Energy Savings[4]

EPC Projects by Sector	
Federal government	28
Provincial/territorial government	1
Municipal government	25
Universities/colleges	8
School boards	35 (mostly multiple locations – project bundling)
Healthcare	55
Industrial	6
Commercial	16
Apartments/condos	98
Other	9
Total	280

involved ESCOs which submitted competitive tenders in response to Requests for Proposals (RFPs).

The Energy Services Association of Canada was created in 2010 to advocate greater use of EPCs across Canada, particularly within governments. Further information on this association and the Canadian EPC market is available in its recent annual magazine and on its website.[5]

POLICIES, PROGRAMS AND LEGAL FRAMEWORK

Federal

The federal government established the FBI program in 1991 to improve energy performance in federal facilities through EPCs. To date, 87 projects have been completed with many involving as many as 17 different buildings. Since its inception, the initiative has attracted an estimated $350 million in private-sector funding through the use of EPCs and generated over $45 million in annual savings. While the program has been operating continuously since inception, activities have not been constant. Eleven new projects were initiated in 1995, but none in 2004 and 2011, one in both 2006 and 2010, and two per year in 2008, 2009, 2011, and 2012. In 2016, competitions were held for six military bases with an understanding that all the remaining 25 bases will have EE retrofits implemented under EPCs over the ensuing five years.

The FBI has a few dedicated staffers that support the different participants in the initiative. It offers energy management training, assistance in raising employee awareness, model RFP and contract documents, as well as step-by-step guides.[3] The FBI also maintains a list of qualified bidders and plans seminars and networking events. Moreover, it offers design/analysis tools and publishes best-practice case studies. Most importantly, it assigns a dedicated FBI Program Officer to support each project.

There are eight FBI case studies available on the FBI website[6] and seven on the Energy Services Association of Canada website.[7]

The FBI office also organizes community-of-practice meetings to share experiences and ideas. The FBI website notes that "this is an innovative networking group that uses the shared experiences of seasoned real property and environmental managers to help federal energy managers develop the best possible EE tactics and strategies".[3]

The Federal Sustainable Development Strategy (FSDS) is another important driver behind the use of EPCs to improve EE in existing federal buildings. In 2008, the Federal Sustainable Development Act was adopted, which establishes the legal framework for developing and implementing the FSDS. The Act requires the development and publication of a sustainable strategy every three years. It also identifies 26 departments and agencies that are responsible for preparing their own sustainable development strategies. Fifteen other federal organizations also voluntarily contribute to the FSDS.

The 2016-19 FSDS is centered on 13 aspirational long-term goals, one of which is low-carbon government. The overall target is to reduce GHG emissions emanating from federal government buildings and fleets by 40% below 2005 levels by 2030, with the aspiration of achieving this by 2025.

Although meeting FSDS targets is a great challenge, one very positive recent development is that the federal government's Office of Greening Government Operations (OGGO) has been recently transferred from the Department of Public Services and Procurement Canada (PSPC) to the Treasury Board. The OGGO works with federal government departments to reduce the carbon footprint of government operations by providing advice and guidance, as well as compiling and reporting results. They have identified FBI as a "key mechanism to help departments achieve their emission reduction targets" established by the FSDS. Moving the OGGO from the PSPC to Treasury Board sends a very clear signal to all departments that the federal government is determined to meet its ambitious targets.

Another interesting recent federal development is the announcement of an additional $120 billion to be spent on infrastructure over the next ten years, with $40 billion specifically earmarked for green infrastructure. While much of this fund will likely be used to construct new buildings, it can also be used to fund upgrades to existing buildings.

Atlantic Canada

Like all other regions of Canada, there have been a wide range of successful EPC projects across the Atlantic Provinces. Case studies have been published for the Canadian Forces Bases at Gander, Halifax, and Gagetown, as well as for Memorial University projects.

None of the four provincial governments in this region has an active program to promote the expanded use of EPCs at this time. There was an initiative in Nova Scotia to select ESCOs to undertake EPCs for public buildings in four regions; although ESCOs were selected, no projects were authorized. One recent interesting development relates to work undertaken by EfficiencyOne, at the request of the Government of Nova Scotia, to investigate the potential use of public purpose energy service companies (PPESCOs). These companies use a similar model as traditional EPCs, with the same delivery process (audit, proposal, contract, financing, installing, commissioning and training), savings guarantees, and ongoing M&V and reporting. The differences are that PPESCO contracts target only public buildings, are smaller than traditional EPCs (i.e. projects less than $1 million), and PPESCOs operate as non-profits, use capital from third-party sources (e.g. social impact bonds or community economic development investment funds) and seek lower rates of return. The Nova Scotia government has expressed interest in working with existing ESCOs to implement one project using this model as a pilot Efficiency Nova Scotia program. If successful, Efficiency Nova Scotia might adopt it as a standard program offering.

Quebec

As noted earlier, Quebec saw the birth and use of EPCs 35 years ago. The Quebec government is recognized as having been the most active and consistent at promoting the use of EPCs for the past 35 years. This has been largely accomplished through direct discussions between provincial ministries responsible for K-12, colleges/universities and hospitals and their related agencies. Those with the highest energy bills were explicitly required to use EPCs to reduce energy costs. It is interesting to note that the by-law entitled Règlement sur les contrats de travaux de construction des organismes publics, chapitre C-65.1, r.5 specifically permits the use of EPCs in Quebec.

One of the more specific features of the process for selecting successful ESCOs in Quebec is the use of Net Present Value (NPV) of proposed projects. This approach involves calculating a total of the discounted annual revenues from a project plus the residual value at the end of the project minus the initial project cost. This approach is used because it encompasses a number of interesting features. Among others, it can simplify the evaluation of one proponent's proposal with those of others. The main weakness of the measure is that the signed contracts do not typically include provisions for failure to achieve proposed savings after the contractual period.

Ontario

Two thirds of all Canadian EPC projects completed over the last ten years have been completed in Ontario. Even when excluding the 94 projects undertaken by one organization (Toronto Community Housing Corporation—[TCHC]), almost 50% of all remaining projects were in Ontario. Moreover, this level of activity has been achieved, for the most part, without any clear leadership or advocacy from the provincial government, its ministries or agencies.

Aside from FBI, the two programs undertaken by TCHC resulted in the largest investment by any other organization in Canada. The Building Renewal Program, which ran from 2005 to 2009, resulted in upgrades to 28 apartment buildings and 33 townhouse blocks, for a total of 6,926 suites. In all, $112 million was invested in building upgrades through the use of EPCs. The subsequent Building Energy Retrofit Program, which ran from 2009 to 2012, resulted in upgrades to a further 26 apartment buildings and five townhouse blocks, for a total of 6,144 suites. Under this program, $57 million was invested in building upgrades through EPCs.

One interesting development that could lead to increased interest in using EPCs is the introduction of Regulation 397/11 under the Green Energy Act that required all municipalities, universities/colleges, school boards, and public hospitals to begin submitting annual reports on their energy usage and resulting GHG emissions as of 2013. The same regulation also required these organizations to submit plans that include proposed measures to reduce energy consumption and GHG emissions as of 2014. Ontario is the only state or province in North America that requires such reports. The province has followed up on this initiative with a requirement that all commercial buildings disclose their energy consumption, starting with buildings over 250,000 sq. ft. in 2018, and then expanding this requirement to all buildings over 50,000 sq. ft. by 2020. Both initiatives are expected to increase interest in seeking opportunities to reduce energy consumption and GHG emissions, some of which could include using EPCs. Figure 15-2 summarizes the estimated GHG emission reductions that could be made by each type of public-sector building, using the GHG emission data that the abovementioned organizations are required to provide. The potentially available 1.6 MT GHG reduction represents over 10% of the gap currently faced by Ontario in meeting its 2020 emission reduction target.

	Emissions (Mt CO²)	40% Savings (MT CO²)
Municipalities	0.988	0.395
Ontario Gov't	0.123	0.049
Federal Gov't	0.232	0.093
Post Secondary	0.654	0.262
Schools	1.488	0.595
Hospitals	0.703	0.281
Total	4.188	1.675

1.7 MT is 27% of 6.4 MT needed to 150 MT target

Figure 15-2: Estimated GHG Emission Reductions Potential by Public Sector. Source: Energy Services Association of Canada[8]

Another development of note in Ontario is the Tower Wise program developed by the Toronto Atmospheric Fund (TAF). Now known as The Atmospheric Fund, this organization was created in 1991 by the City of Toronto with a $23 million endowment to help it achieve energy/GHG reduction targets. Tower Wise focuses on encouraging owners of rental apartments, condos or social housing to upgrade EE in their buildings by using an EPC. These are typically smaller projects which are often too small for traditional ESCOs to provide performance guarantees. Instead, TAF links owners with Energi, a private company that offers an insurance product to secure energy performance guarantees for a fee, typically 2 to 4% of the project value. There are seven case studies on the TAF website, http://towerwise.ca/, consisting of three condos, two social housing units and two rental apartments.

The most recent development in Ontario is the Climate Change Action Plan. Released in 2016, it summarizes how the province intends to use the $8 billion in revenues expected to be generated under its new cap-and-trade program for pricing carbon emissions. This plan includes $380 to $500 million to retrofit social housing, $400 to $800 million to retrofit schools, hospitals and universities/colleges and $90 to $100 million for EE retrofits of its own buildings.[9] One of the recommendations in this Plan is that the government enable the use of EPCs across the OPS, the Ontario Public Service which consists of all government ministries.

Manitoba

Manitoba has been actively promoting EE for many years, mainly through programs managed by Manitoba Hydro. Although the government or Manitoba Hydro did not actively promote EPC projects in the past, there have been a few successful EPC projects.

This could change as the government is in the process of creating Efficiency Manitoba, a new Crown Corporation that will be mandated to achieve electricity savings of 1.5% and natural gas savings of 0.75% per year for the next 15 years. Using EPC to achieve these objectives is among the opportunities that this new agency may consider.

Saskatchewan

There were 13 EPC projects in Saskatchewan in the last 10 years, almost as many as in the other three western provinces (Alberta, British Columbia and Manitoba) combined. The main reason for this was that SaskPower, the provincially owned integrated electricity utility, actively promoted the use of EPCs by its institutional customers. They also sought an ESCO partner to undertake projects and, after a competitive bid, formed a joint venture with Honeywell. Their five-year agreement was renewed for a further five-year term. Like Alberta, Saskatchewan offers long-term capital to public enterprises through a debenture program.

Alberta

Alberta Infrastructure managed the most active and successful program on a relative basis to promote the use of EPCs in provincial buildings of any government to date in Canada. Initiated in 1995, retrofits were completed in over 150 facilities with various vendors for projects totaling $28 million. These projects contributed an estimated 10% reduction in energy use over the decade that the program ran.[10]

Another leading innovation in Alberta was the Capital Borrowing Regulation under the School Act. Initially passed in 1988 and later amended, it requires that if school boards borrow funds to retrofit a school to reduce energy consumption, the service provider must offer a performance guarantee. Recently, the Presidents of the Alberta School Boards Association (ASBA), College of Alberta School Superintendents (CASS) and the Association of School Business Officials of Alberta (AS-BOA) issued a joint letter to all School Board Chairs, Superintendents and Secretary-Treasurers encouraging them to investigate the potential of using EPCs for schools in their boards. One of the leading boards that uses EPCs is the Edmonton School Board.

The recent final report from the Alberta Energy Efficiency Advisory Panel, entitled "Getting it Right: A More Energy Efficient Alberta," included among its recommendations that the Alberta government consider expanding this mechanism to other institutions. It also added that a complementary action would be for the government to formally authorize the use of EPCs for public-sector buildings.[11]

Alberta Health Services recently entered into EPC contracts for the Alberta Hospital and the Royal Alexandra Hospitals in Edmonton, with potential plans to use these contracts at other hospitals throughout Alberta.

British Columbia

The Green Buildings BC Retrofit Program was launched in 1996 and designed to promote energy reduction in B.C.'s provincial building stock. The program was successful in assisting school districts, universities and colleges in undertaking energy reduction programs.

The second-largest EPC initiative in Canada was undertaken by BC Housing. From 2009 to 2012, energy and infrastructure improvements were made in 5,000 social housing residences in over 300 buildings; total project cost was $120 million. In addition to saving $3.3 million per year, GHG emissions were reduced by 5,000 tons with significant reduction in deferred maintenance backlog.

B.C.'s carbon neutral government program is legislated under the Greenhouse Gas Reduction Targets Act and the Carbon Neutral Government Regulation. It requires all public-sector organizations to follow a five-step process to achieve carbon neutrality. In their initial Energy Plan, it was estimated that $1.5 billion in EE upgrades were required.

Facilitators

As noted in the sections above, FBI and the Quebec government encourage public entities to employ independent facilitators to assist them with projects. This has been a particularly important success factor when project managers were unfamiliar with EPCs.

Some ESCO clients have found this service to be critical. On the FBI website, the following quote is made by Karen Dupuis of the RCMP Northwest Region: "If the (FBI) facilitation services were not available, I don't think we could have moved forward with this project. It just wouldn't have gotten off the ground." A recent article provides some examples of the important roles that facilitators can play.[12]

CONCLUSIONS

At the federal level, it is encouraging that the recent federal budget allocated an additional $13.5 million to the FBI program, that the

Department of National Defence has signed its intention to use EPCs to undertake major EE retrofits at every military base in Canada, and that the Treasury Board is now responsible for the Office of Greening Government Operations and the achievement of FSDS targets.

At the provincial level, the Energy Services Association of Canada has identified the following five

based on a review of best practices in other jurisdictions:[8]

1. **Authorization that EPCs can be used by public-sector buildings**—Although EPCs have been successfully used in every province, no provincial government has publicly acknowledged that government departments and the public sectors they control (Broader Public Sector or BPS) can use these contracts. The U.S.-based National Association of State Energy Officials (NASEO) recently noted in a report that every state in the US has provided such authorization. Such authorization is also clear at the federal level in Canada and is promoted by FBI.

2. **Encourage governments/BPS to use EPCs**—This is not only common in every state in the U.S., but also at the federal level in Canada through FBI.

3. **Identify lead management agency to promote use of EPCs**—Provincial governments should follow the lead of federal government in this regard as well. In the U.S., the NASEO report also notes that every state in the U.S. has identified a lead agency to promote broader use of EPCs. The NRCan FBI program has this responsibility for all government federal departments.

4. **Empower lead agency with staff to promote EPCs**—The U.S.-based Energy Services Coalition has identified six leading states where lead agencies assist with financing as well as general support for EPCs. This model should be duplicated in Canada.

5. **Use EPCs to provide funding to match federal programs**—As noted earlier, there is currently a particular opportunity to use EPCs to provide the provincial portion of the matching grants for EE retrofits to green infrastructure and social housing.

Chapter 16

Program Evaluation

Ms. Marie COUTURE-ROY, Econoler

The growth of EE in Canada is largely driven by the various ratepayer-funded programs put in place by governments, utilities and non-profit organizations. With the ultimate goals of shifting the market toward more EE technologies and transforming behaviors, these programs represent one of the most successful and cost-effective solutions to reducing energy usage and fighting climate change.

Achieving these objectives requires regular program evaluations to continuously improve impacts on the market and ensure programs are on track to cost-effectively achieving long-term objectives. Hence, most governments, utilities and non-profit organizations that manage EE programs in North America conduct regular evaluations.

WHAT IS EE PROGRAM EVALUATION?

EE program evaluation is the process of determining and documenting the results, benefits, and lessons learned from an EE program. The objective is not to penalize stakeholders involved in evaluating programs (designer, implementer, funder, etc.), but rather to promote continual improvements in program delivery.

Program evaluation is conducted through a credible, transparent and independent process in real time and/or retrospectively. It relies on methodologies and protocols developed by the most active jurisdictions in the field of EE, mainly in the U.S., such as the California Energy Efficiency Protocols, the Office of Energy Efficiency and Renewable Energy guide for managing general program evaluation studies for the evaluation of DSM programs, the National Energy Efficiency Evaluation, Measurement and Verification Standard of the Lawrence Berkeley National Laboratory, the Energy Efficiency Program Impact Evaluation Guide prepared by the State and Local Energy Efficiency Action Network (SEE Action), and the more recent protocols of the Uniform Methods Project (UMP) developed by the U.S. Department of Energy (DOE). These approaches are continuously being improved, mainly based on knowledge gained from the International Energy Program Evaluation Conference (IEPEC) which is held every two years in the U.S. to keep EE program planners, implementers and evaluators informed on the latest publications in the field.

EE program evaluation also allows organizations implementing DSM programs to report the real impacts and market influences to relevant authority (regulators, executive boards, stakeholders, taxpayers, etc.). Essential components of the DSM program cycle include ensuring that programs run per program theory, monitoring program performance, tracking market evolution, and finally determining whether programs are still needed.

APPROACHES

Mainly three evaluation approaches are used in Canada:
- Process evaluation serves to analyze how efficiently a program is delivered with respect to objectives. Process evaluation also allows understanding the design of a program and answers the following questions: (i) Was the program theory and associated logic diagram correct? (ii) Is the market responding to the program as expected?

- Market evaluation serves to measure and improve program influence on the market. Market evaluation includes data-collection activities to establish program impacts and evaluate the level of technology penetration to assess remaining potential. This process permits improving program impacts based on participant and non-participant feedback.

- Impact evaluation focuses on determining program gross and net savings, GHG reductions and other non-energy benefits to assess program cost effectiveness. The aim of impact evaluation is to compare program objectives to savings targets and program results. These results are achieved as follows:
  — A gross impact evaluation is developed through a combination

of billing analyses, building simulations, engineering methods, onsite visits, and measurement and verification (M&V) based on the International Performance Measurement and Verification Protocol (IPMVP).
— Net impacts are established by adjusting gross impact to account for market effects, free riders and program-induced spillover.
— Non-energy benefits are also evaluated and include economic impacts and environmental benefits.

SPECIFIC APPLICATIONS

EE program evaluation consists of designing an adapted evaluation plan that should be developed during the program design phase and should include at least the following components: (i) the types and scope of evaluations; (ii) the timeline of evaluation activities; and (iii) evaluation methodologies (including measurement strategies).

LEVEL OF INVESTMENT

The typical cost of an EE program evaluation usually varies between 3 to 5% of the total program budget. The final cost depends on the frequency, complexity, required level of certainty and scope of data collection and analysis.

MARKET ACTORS

Most EE program evaluations in Canada are conducted by third parties to ensure unbiased and credible results. Provincial energy regulators require that most organizations, such as the IESO and Efficiency Nova Scotia, conduct annual independent evaluations. Others carry out a mix of internal and external evaluations. For internal evaluations, evaluation reports are often peer reviewed by an evaluation advisor.

Table 16-1 presents examples of provincial utilities and Canadian organizations that evaluate their DSM programs and how (internal, external) they accomplish this.

Table 16-1

Province	Utility/Organization	Internal Evaluation	External Evaluation
Alberta	Energy Efficiency Alberta		X (planned)
British Columbia	BC Hydro	X	X
Manitoba	Manitoba Hydro	X	X
Ontario	IESO		X
	Ontario Energy Board		X
Québec	Hydro-Québec		X
New Brunswick	NB Power	X	X
Nova Scotia	Efficiency Nova Scotia		X
Newfoundland and Labrador	Newfoundland Power and Newfoundland and Labrador Hydro	X	X
Yukon	Yukon Energy Corporation	X	

Complete program evaluation, including process, market and impact assessments, requires a team comprised of expert engineers, market researchers and statisticians with solid knowledge of the EE field, program design and implementation, as well as regulatory processes. Only a few Canadian firms have the high level of expertise and skill to carry out EE program evaluations, which explains why U.S. based firms are occasionally granted these mandates. Still, the tendency is to favor the use of Canadian firms who now have the equivalent experience as their American counterparts.

Most evaluation assignments are granted through a public request for proposals (RFP). RFPs often require the use of local market research and engineering firms for field work such as surveys and site verification studies to foster the development of local capacity.

Chapter 17

Capacity Building and Training

Mr. Mathieu CÔTE, Canadian Institute of Energy Training (CIET)

The lack of awareness, knowledge and capacity in the EE field has long been identified as one of the main barriers to large-scale adoption of EE in Canada. As far back as the late 1980s, Natural Resources Canada (NRCan) in its previous incarnation as Energy, Mines and Resources Canada commissioned the creation of a series of practical manuals known as the Energy Management Series which dealt with the principles of EE in a number of generic energy-consuming systems and technologies. The manuals also included a methodology for managing energy. To enhance the use of this library of information in the marketplace, NRCan developed a college-level training program known as CEMET (Canadian Energy Management and Environmental Training) whose curriculum was based on the Energy Management Series. Durham College, in Oshawa, ON, was appointed as the lead college for this pilot project conceived as a precursor to establishing a national college training network. After CEMET had been operating for a couple of years, it became apparent that the college model was not ideally suited to program objectives. Furthermore, an identifiable marketplace delivery organization was needed if the program was going to penetrate the emerging energy management sector. Therefore, while CEMET remained the original NRCan program, a privately-owned company was created to serve as the delivery agent.

In 1996, NRCan terminated its CEMET contribution agreement. Energy management in those days was a hard sell, particularly since the 1980s oil price shock had faded. In 1997, NRCan developed and launched the Dollars to $ense (D2$) training curriculum. Since its inception, over 30,000 representatives from industrial, commercial and institutional organizations throughout Canada have enrolled in the D2$

workshops, making them by far the most successful EE training program in Canada.

In response to market demand, other organizations have started
offering EE and energy management training. However, the sheer number of participants in D2$ and the foundational nature of the workshops
clearly attest that NRCan opened up the market for those organizations
to successfully offer structured and sustainable energy training. Given
that NRCan licensed the D2$ workshops to a private organization in
2016, it is safe to assume that a market transformation has occurred in
EE and energy management training. The public sector has mostly pioneered and created the first sustained offerings in Canada, paving the
way for the private sector to develop and offer more courses on more
topics in more places.

This same process could occur in the near future for training related
to net-zero or near-net-zero construction for example. The private sector
is unlikely to invest large amounts of money to develop comprehensive
professional training or certification programs until there is high enough
demand for these types of courses. However, public institutions can accelerate market development by being the catalyst in the field. They can
also foster the development of cutting-edge fields, politically desirable
areas or policies related to climate change or energy.

CAPACITY BUILDING AND
TRAINING ACTIVITIES

To develop sustainably, the EE sector requires a diverse range
of actors from policymakers to building operators. Unfortunately, not
enough Canadian college or university degrees as well as professional
schools and institutes cover EE. EE training is thus largely the subject
of continuing education, but offer and demand are growing quickly in
this sector. Professional designations are widely recognized and sought
after. The current Canadian EE training ecosystem is the result of a long
process that began in the 1980s, in which all types of organizations have
been involved in one way or another.

One of the most prevalent forms of EE training in Canada involves
energy management. Energy management is a niche of EE, which focuses more on how to save energy from existing equipment, facilities
and installations rather than how new equipment could be designed to

be more energy efficient or implemented in an energy efficient manner. Energy management is an art and full-time energy managers continuously identify energy conservation measures, thereby annually saving their organizations hundreds of thousands of dollars, if not more, and justifying their own salary in the process.

Several highly established energy management certifications are currently offered across the country, including the following:

Table 17-1: Some Energy Management Certifications Currently Offered Across the Country

Program	Topic	Certification Body
Building Operator Certification (BOC)	Sustainable operation and maintenance	Northwest Energy Efficiency Council (NEEC)
Certified Commissioning Professional (CCP)	Building commissioning	Building Commissioning Association (BCxA)
Certified Energy Manager (CEM)	Energy management	Association of Energy Engineers (AEE)
Certified Measurement and Verification Professional (CMVP)	Measurement and verification	AEE/Efficiency Valuation Organization (EVO)
LEED Green Associate/LEED AP	Green building design, construction and operation	Canada Green Building Council (CaGBC)

Each of these programs, and others, covers a specific aspect and/or a specific range of actors in the EE sector. For example, the CEM is a comprehensive five-day certification program for energy engineers and energy managers. A throng of shorter, non-certifying energy management courses are also offered by a wide array of for-profit and not-for-profit training providers, as well as colleges and universities.

MAIN STAKEHOLDERS

Indeed, the most defining aspect of the Canadian EE training ecosystem is the various types of involvement of diverse organizations from utilities to private companies.

In the private sector, the Canadian Institute for Energy Training (CIET) has been offering leading EE training and certification programs since 1996. Having trained close to 10,000 individuals as of 2017, CIET has established itself as a reference in Canadian EE training. Several other for profit and not-for-profit organizations offer

EE training, such as ASHRAE chapters, the Canada Green Building Council, several community colleges with established EE curricula, as well as the Building Owners and Managers Association of Canada (BOMA), just to name a few.

EXAMPLES

Federal
Through NRCan, the federal government has developed and offered several EE training programs over the years. The most recognized is certainly the aforementioned D2$ workshops. At various times, NRCan has offered training on software or tools (RETScreen, CAN-QUEST, ENERGY STAR® Portfolio Manager) and developed the four-day Advanced Building Recommissioning (RCx) program, to name a few. By providing this type of training, NRCan plays a key role in helping EE professionals do their job more effectively and efficiently.

RETScreen is a clean energy management software system for EE, renewable energy, cogeneration project feasibility analyses as well as continuous energy performance analyses. It aims to empower professionals and decision makers to rapidly identify, assess and optimize the technical and financial viability of potential clean energy projects. RETScreen allows managers to easily measure and verify the actual performance of their facilities and helps identify additional energy savings or production opportunities.

CAN-QUEST is a Canadian adaptation of eQUEST, the popular U.S. building energy simulation software. It is used to demonstrate performance path compliance with the NECB 2011 and support the design of high performance commercial and institutional buildings.

The ENERGY STAR® Portfolio Manager (ESPM) is the U.S. Environmental Protection Agency's free interactive energy management tool which enables tracking and assessing energy and water consumption across an entire portfolio of buildings. It offers weather-normalized energy use intensity values, GHG emission metrics, reporting features that help track trends over time, and 1-100 energy performance scores for eligible building types. NRCan has adapted the ESPM to the Canadian market. It is used as a standard energy benchmarking tool that provides accurate and equitable energy assessments for the Canadian commercial and institutional buildings sectors. It should be noted, however, that buildings are compared to others based on unverified data, so results are not always accurate.

While identifying and measuring the impacts of EE training is tricky, several studies conducted on federal training initiatives have clearly demonstrated that it is a sound investment. For example, a study conducted by Habart & Associates for NRCan[i] demonstrated that the Dollars to $ense Workshops save 225 kt of CO_2 per year.

Provincial

Utilities have assumed a leadership role in EE programs in recent years, and the pressure placed upon them to reduce energy consumption means they are among the first to recognize the need for a skilled and trained workforce.

One key example of provincial involvement in EE training is Ontario's IESO. It operates what is potentially the most successful energy management training incentive program in Canada through the Save on Energy brand. From 2011 to 2015, more than 6,000 individuals underwent IESO-incented training. IESO expanded its range of programs eligible for funding in 2017 and now maintains a target of training 10,000 individuals from 2017 to the end of the Conservation First Framework in 2020. This certainly represents a bold and significant commitment to increasing the capacity of EE stakeholders to generate, accelerate and replicate EE projects throughout the province. IESO currently provides incentives for a vast array of training covering the following subjects: (1) foundational development; (2) specialized development; and (3) professional development.

Other utilities, governmental bodies and private organizations recognize, require or support numerous EE certification programs, from BC Hydro to FortisBC, the Government of Alberta, the City of Toronto and Énergir, among others.

---

[i] Impact Attribution for Dollars to $ense Workshops, Habart & Associates, 2003.

Chapter 18

The Canadian Energy Efficiency Alliance (CEEA)

Mr. Martin LUYMES
Heating, Refrigeration and Air Conditioning Institute of Canada (HRAI)

INTRODUCTION

The Canadian Energy Efficiency Alliance (CEEA) is Canada's leading independent advocate promoting the economic and environmental benefits of EE. Established in 1995 to not only increase competitiveness and environmental protections, but also improve how stakeholders collaborate to promote EE in Canada, this not-for-profit organization works with federal and provincial governments as well as stakeholders to ensure EE is a priority for all sectors of the economy. By monitoring, examining, and developing energy-efficient public policy ideas, programs, and standards, CEEA is an effective resource for policymakers, businesses, consumers, energy companies, and environmental groups.

CEEA members include large international corporations, Canadian utilities and a variety of industry associations with an interest in EE. CEEA counts among its members the Canadian Construction Association, Electro-Federation Canada (EFC), the Heating, Refrigeration and Air Conditioning Institute of Canada (HRAI), and the North American Insulation Manufacturers Association (NAIMA), collectively representing more than 25,000 like-minded industry members.

CEEA main services have included: advancing and facilitating research on EE issues; creating networking opportunities for members and stakeholders; supporting effective EE policies, programs, codes, and standards; helping members develop, promote, and deliver energy efficient products and services; and generally raising awareness about EE across all sectors of the Canadian economy.

In the last few years, the CEEA has become a force to be reckoned with in Ottawa, having gained the attention of regulators and policymakers in key government departments and agencies. The push to address climate change, internationally and at the federal and provincial/territorial levels in Canada, has breathed new life into the EE movement and created compelling new opportunities for CEEA and its members. Investing in EE has become a key policy priority in Ottawa and a central focus in a number of provinces, as exemplified most visibly by the establishment of new EE-related Crown Corporations in Alberta, Manitoba, Ontario and Quebec.

In 2016, the CEEA Board of Directors introduced a plan to take the organization to the next level: CEEA, Version 3.0. The new CEEA will build on past successes while breaking new ground. Before outlining some of the key elements of this plan, it might be helpful to recount some of the history of CEEA. To see where we are going, it is helpful to recall where we came from.

THE ORIGINS OF CEEA

In 1993, a diverse group of individuals and organizations representing key EE interests in the Province of Ontario formed the Energy Efficiency Consortium. The core mission of the group was "to rapidly accelerate the implementation of EE in Ontario buildings, equipment and processes." An initial report prepared by the consortium identified five major barriers to EE, seven common mitigation strategies and 30 specific actions, the first being to establish the Canadian Energy Efficiency Alliance to act as the vehicle for continuing the work of the Consortium.[i] In 1995, CEEA was born.

The initial mandate of CEEA was "to provide leadership in promoting EE as a means of achieving economic and environmental benefits." Bruce Lourie served as the first Executive Director. In its first year, CEEA's submission to the Macdonald Advisory Committee on competition in Ontario's electricity system recommended the creation of a rate-

---

[i] It is worth noting that many of the 30 recommendations of the Energy Efficiency Consortium report continue to be relevant today and may well inform the future direction of a revitalized CEEA. As Michael Lio, an early proponent of CEEA, states, "beyond the value of the recommendations issued, the Consortium and the Canadian Energy Efficiency Alliance that came out of it demonstrated that stakeholders can agree on the challenges that Canada faces and solutions to those challenges."

payer-funded EE fund which was, in fact, later introduced and became the source of funding electricity conservation programs in Ontario.

A few years later, CEEA launched its first successful advocacy campaign to retain the EE provisions in the Ontario Building Code that were slated to be removed by the then newly-elected government. This success was shortly followed by the creation of EnerQuality in 1998, a for-profit company formed by CEEA jointly with the Ontario Home Builders' Association (OHBA) to manage the R-2000 program and subsequently the EnergyStar labeling program for new homes. EnerQuality programs were designed by and for builders to suit the realities of the homebuilding sector while highlighting important values such as responsible energy use and enhanced consumer protections.

In the early 2000s, CEEA began publishing its Energy Efficiency Report Cards, an idea pioneered by the American Council for an Energy Efficiency Economy (ACEEE). During this period, Peter Love led the organization as Executive Director. The Alliance worked to expand membership across the country but remained especially active in Ontario, among other things advocating the establishment of province-wide funding for EE programs. These efforts led to the creation, in 2004, of the Ontario Power Authority which was authorized to spend $350 million per year on electricity conservation programs.

CEEA VERSION 2.0

By 2011, the CEEA Board of Directors had come to the conclusion that the organization was suffering a malaise and in need of a reboot. While a number of successful programs had been launched and advocacy goals achieved, there was an identified need for more visibility at the federal level and a higher level of engagement and communication with key industry stakeholders.

To make the shift to more proactive government relations and advocacy work on behalf of its members, the organization hired a new President and CEO, Elizabeth McDonald. A veteran with a strong track record of leading organizations that served industries undergoing dramatic change (i.e. CanSIA and the solar industry), Ms. McDonald brought the leadership skill set that CEEA sought: an ability to not only discern industry challenges, but also identify and implement solutions through active collaboration with governments and key stakeholder groups.

Under Elizabeth's leadership, CEEA more than doubled membership, the board became more diversified with representatives from coast to coast, and CEEA has now become both a focal point for many EE associations and an effective voice for the business of EE. The organization introduced a variety of valuable new programs and services, including a series of research projects on consumer and business attitudes toward EE, a popular blog and, more recently, a new semi-annual publication entitled Efficiency Matters, as well as a very popular and cost-effective series of lobbying Days on the Hill events in Ottawa.

These programs, along with the CEO's tireless efforts and focused action on a variety of key policy areas, elevated the status and visibility of CEEA in federal public policy circles. In so doing, however, the organization also had the effect of elevating expectations among the membership, partners, stakeholders and government. These expectations occasionally went well beyond what the CEEA was able to deliver with its limited resources.

In the context of newly-elected governments at the federal and provincial levels that were suddenly friendly to the idea of EE as a primary policy goal, especially in the transition to a low-carbon economy, CEEA was once again at a crossroads. The confounding but exciting new challenge became: How can we muster the wherewithal to step up and lead this sector forward so that the full promise of EE can be realized?

THE MOVE TO CEEA VERSION 3.0

In 2016, the CEEA Board and a few friends of the organization held a strategic planning discussion that focused on building a new more sustainable version of CEEA. The board declared that it was pleased with the progress made under the leadership of Elizabeth McDonald, having realized much of the potential that CEEA Version 2.0 promised less than five years prior. However, the prospect of financial sustainability remained elusive. How could CEEA fully lead the Canadian EE sector forward unless it dramatically increased access to financial resources?

Among other conclusions, the board decided it should look into partnerships with like-minded family foundations. What followed was close to a year of discussions with the Clean Economy Fund (CEF), a relatively new funders' collaborative spearheaded by the Ivey Foundation and designed to maximize and leverage Canadian philanthropic grants in support of a clean, low-carbon and prosperous Canadian economy.

To aid their discussion, CEEA and CEF conducted a market needs assessment to determine what would be needed to accelerate EE investment in Canada. The number one recommendation was that Canada needed "One National Voice for Energy Efficiency," with a focus on collaboration and cooperation among sectors, as well as on educating consumers and policymakers. CEF and its funding partners further concluded that the logical choice to serve as this "one national voice" should be CEEA.

Accordingly, CEEA collaborated with CEF, and subsequently Carleton University in Ottawa, to design and build a new more viable multi-sector, non-profit organization that would achieve measurable and sustained EE gains in the next decade.

There is much work yet to be accomplished to craft a new vision for CEEA, and this work will be left to new leadership, but the vision will include at least the following elements:

- Making EE top of mind for key decision makers, to the point where it is viewed as a central means to increasing economic productivity, creating jobs, and reducing GHG emissions;

- Improving Canada's international EE ranking;

- Promoting deep retrofits of the existing housing and building stock with clear targets for reducing energy consumption thereof;

- Pursuing national commitments to net-zero energy in new constructions.

CEEA has already concluded a search for new leadership, and the new CEO has been in place since early 2018. These changes are necessary, indeed essential, to allow the organization to once again live up to the promise of its vision and mandate. CEEA and its partners are setting the table for a renewed and reinvigorated organization that will have the resources and industry support to truly advance the cause of EE in a meaningful way across the country.

Thanks to Peter Love, Bruce Lourie and Michael Lio for their contributions to this initiative and for their vision, leadership and action in the early days of the Alliance, and, of course, to Elizabeth McDonald for breathing life into the organization and keeping the original vision alive during her five years as the face of CEEA.

Chapter 19

EfficiencyOne: A Unique Implementation Model

INTRODUCTION

Nova Scotia's EfficiencyOne is unlike any other EE provider in Canada: it is the country's first and only independent EE utility. Neither a branch of government, nor an electrical utility department, it operates independently and is acknowledged as an innovative leader for its performance and governance achievements.

EfficiencyOne is a leader in the design and delivery of resource efficiency programs and services for households, businesses, and large industrials, with a customer satisfaction rating that consistently exceeds 90%. Federally-incorporated and led by an independent board of directors, EfficiencyOne is based in Halifax, Nova Scotia.

Since 2010, EfficiencyOne (formerly Efficiency Nova Scotia Corporation) has evolved from a startup operation to a North American industry leader and Canada's only EE utility that works with over 200 contractors and partners and provides services to all electricity customers across Nova Scotia. The organization also has a contract with the Province of Nova Scotia to deliver an EE program for low-income Nova Scotians who heat their homes with oil and other fuels.

Through Efficiency Nova Scotia, EfficiencyOne manages a portfolio of 25 programs and pilots that help Nova Scotian clients in all market sectors become more efficient and reduce energy consumption. These programs offer financial support and expertise from home weatherization to industrial process upgrades. Through Efficiency Nova Scotia programs, clients are able to implement projects that meet their investment requirements, with simple paybacks sometimes as short as a few months.

EfficiencyOne has a proven track record of achieving targets within budget, and all programs undergo a rigorous quality assurance process. Its energy and demand savings programs are evaluated annually by independent third parties and verified by the Utility and Review Board (UARB) using the latest industry best practices. It focuses on continuous improvements and innovations to ensure programs and services lead the energy sector while remaining highly responsive to the needs of customers. These proven program management capabilities are complemented by solid expertise in research, regulatory affairs, marketing, outreach and education, as well as stakeholder engagement.

With an exemplary culture that fosters innovation and rigorous focus on measurable results, EfficiencyOne adopts and often establishes industry best practices in all programs. It supports its programs through expertise in research, technology, quality control, capacity building, marketing and public education. All programs and processes are thoroughly documented and managed using robust administrative systems. Annual independent evaluations objectively measure energy savings and drive continuous improvement. This structure is adaptable to other jurisdictions, making it possible to rapidly deploy state-of-the-art programs in new markets.

RESULTS

As a result of the EE programs operating in Nova Scotia since 2008, electricity users now save more than $123 million each year. These savings will continue to rise and by the end of 2016 corresponded to 9% of annual utility revenue from electricity sales, as demonstrated in Figure 19-1.

Achieving substantial energy savings in any jurisdiction requires developing the local economy. Nova Scotia has grown its EE industry capacity through contractors, consultants and equipment suppliers. In a province with a population of approximately one million, the efficiency industry employs 1,200 professionals and hundreds of businesses. EE projects account for more than $180 million in annual private-sector investment. The efficiency sector is growing at an annual rate of 8%, compared to annual provincial GNP growth of 1.5%.

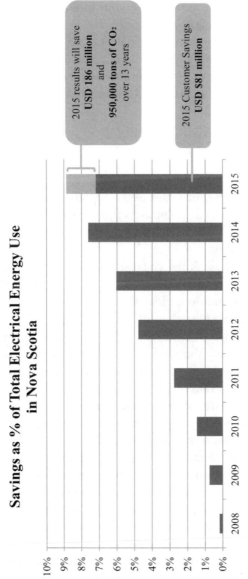

Figure 19-1: Electricity Savings Due to EE Programs Operating in Nova Scotia

SERVICES

The unique EfficiencyOne model allows the utility to market its expertise and services to governments, utilities and private-sector customers outside of Nova Scotia.

Creating a new framework for national EE programs and securing support from stakeholders are challenging ventures. Since EfficiencyOne is well-positioned to design and deliver services in Nova Scotia, it now helps other jurisdictions replicate its model and achieve similar successes on an ad hoc basis.

PILLARS OF SUCCESS

To understand the successes of the organization, it is important to understand a number of key elements that bolster EfficiencyOne.

Engaged Stakeholders

First and foremost is the constantly evolving role of local stakeholders. In the consultations that led to the establishment of EfficiencyOne, stakeholders made it clear that they were uncomfortable with having the province's electrical utility in charge of the EE portfolio. There existed an inherent conflict of interest. After all, they reasoned, how could the company that makes money by providing power also be running programs that fostered less and less electricity consumption?

All the while, stakeholders wanted to abandon the government-run model. The decision to go with an independent not-for-profit organization was well received from the beginning. Stakeholders felt they had been listened to and empowered.

It provided a fresh start to the people of the province, as well as a homegrown organization with a trailblazing model and commitment to improving not only the lives of Nova Scotians, but also the economy and environment. It also allowed the team at EfficiencyOne to build an independent organization driven by ideals, led by a board of directors, and regulated by the province's UARB.

This open collaborative model has thrived. Not every stakeholder group agrees on every issue, but they all share respect, are credible, and believe they are contributing to a responsive and responsible EE model.

Talented Workforce

EfficiencyOne has thrived because of its ability to recruit, develop and retain talent. In a province that has at times struggled to retain local youth and attract newcomers, EfficiencyOne has flipped the script. The staff of about 90 people are driven by a core of young, talented, and smart people committed to the innovative field of EE that is an environmentally and economically transformative business model. Another 200 trade partners further support the programs and services offered by the Efficiency Nova Scotia franchise (efficiencyns. ca). This franchise is Canada's first regulated EE utility offering a suite of energy conservation services to residential, commercial and industrial clients.

Commitment of Nova Scotians

The enthusiasm of Nova Scotians for EE has reinforced the model. Indeed, EfficiencyOne stresses the leadership of Nova Scotians themselves in building an EE culture across the province rather than focusing on celebrating the organization itself as the leader.

Support and Leadership from Government and Power Utility

It is important to note the significant role of the province's electrical utility, Nova Scotia Power. The utility recognized the importance and benefits of EE and conservation and brought forward the idea of increased DSM a decade ago. Government, too, has played a significant role in shaping and supporting Nova Scotia Power. From drafting the legislation to create EfficiencyOne, to funding programs for fuels other than electricity, the provincial government has been and continues to be a champion of EE.

Corporate Reputation

The commitment of EfficiencyOne to ensuring the reliability and veracity of energy savings targets further enhances its credibility. Its multistage process begins with EfficiencyOne staff tracking savings throughout the year. Then, an independent evaluator examines the savings records and produces a public report presenting the findings. Finally, the UARB hires a third-party expert to review and verify the report, thus rendering the whole process fully transparent.

CONCLUSION

EfficiencyOne leads by example as a unique organization in Canada. As created, the EE utility model is unique in that it demonstrates how proper innovation to support the development of a sustainable EE market leads to impressive results.

Chapter 20

RETScreen: Clean Energy Management Software

Dinesh S. Parakh and Gregory J. Leng
RETScreen International, CanmetENERGY

BACKGROUND

How does one transition from the objective of clean energy deployment (including EE, renewable energy, and cogeneration) to the practical implementation of cost-effective projects that actually reduce conventional energy consumption, save money, and reduce GHG emissions? That is, how does one identify and implement projects that advance clean energy goals that are also financially attractive?

The RETScreen Clean Energy Management Software arose from a simple observation: the time and cost required to assess various energy options was excessively and often prohibitively high. RETScreen was conceptualized as a means to address the assessment cost barrier that precluded the accelerated implementation of cost-effective projects. RETScreen helps to clearly, concretely, and very inexpensively identify the large number of clean energy projects that are inherently profitable (i.e. profitable even in the absence of incentives or regulations, etc.), thereby eliminating said assessment cost barrier.

Since its first release in 1998, the RETScreen Software has evolved into the world's leading software as measured by metrics such as number of users, built clean energy projects, user savings, and avoided GHG emissions. The software enables professionals and decision makers to rapidly identify and assess the viability of potential clean energy projects, as well as easily measure and verify the actual and ongoing energy performance of buildings, factories, and power plants around the world. By enabling comprehensive full project life-cycle analysis, RETScreen assists in framing the critical business case for engaging in various EE measures. The software has been tested and validated in the marketplace as a tool for sound analysis upon which investment decisions can be based and capital allocated. RETScreen has been continuously improved over the years and the latest generation of the software, RETScreen Expert, was released to the public in September 2016.

Figure 20-1:
Clean Energy Project Life Cycle as Modeled in RETScreen Expert

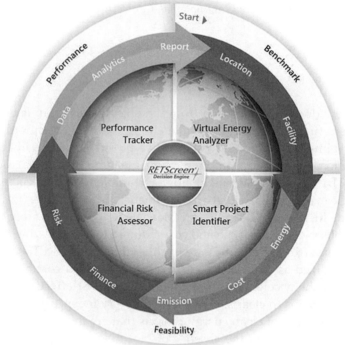

STAKEHOLDERS

RETScreen is developed by the CanmetENERGY Research Centre in Varennes, QC, a national energy laboratory under the auspices of NR-Can. NRCan is the primary source of funding for RETScreen. Additional funding is contributed by national and international sources and task share partners, including the National Aeronautics and Space Administration (NASA), Ontario's IESO, the Vienna-based Renewable Energy and Energy Efficiency Partnership (REEEP), the World Bank, the Global

Environment Facility (GEF), and the United Nations Environment Programme (UNEP).

FEATURES

RETScreen Expert is used to evaluate various types of EE measures for projects in buildings and factories. It can model and analyze the viability of EE improvements in a wide range of residential, commercial, institutional, and industrial buildings and facilities, from single-family homes and apartment complexes, to office buildings, hospitals, as well as large pulp and paper mills. The software is also used to assess projects incorporating a variety of EE measures associated with building envelopes, ventilation, lighting, electrical equipment, hot water, pumps, fans, motors, process electricity, process heat, process steam, steam loss, heat recovery, compressed air, refrigeration, and more. It is useful for both new constructions and retrofits. Whole facilities can be modeled, or sub-systems and rooms can be studied separately. To facilitate rapid analysis, RETScreen also contains fully integrated product, project, benchmark, cost, hydrology, and climate databases (the latter includes 6,700 ground-station locations in addition to NASA satellite data covering the entire world).

As of mid-2017, RETScreen has over 550,000 users worldwide, including more than 100,000 users in every Canadian province and territory. It has been adopted by all types of private industry, various levels of government, the broader public sector, utilities, and is used for teaching and research in the majority of universities and colleges in Canada. Bolstered by a proven track record and reputation for reliability, RETScreen plays an important role in the Canadian EE market as a tool of choice to identify, analyze, and implement viable EE projects.

RETScreen has been an important driver of EE growth in Canada. RETScreen is projected to have over one million users worldwide within the next ten years, resulting in an estimated $20 billion in direct user cost savings, $100 billion in project investments, and 50 million tons/year in GHG emission reductions. Extrapolating for Canada, it is anticipated that by 2022, RETScreen will have approximately 200,000 users, $4 billion in direct user cost savings, $20 billion in project investments, and 10 million tons/year in GHG emission reductions. A large number of these projects will be EE projects, thus confirming RETScreen's status as a practical tool to actualize EE projects in Canada.

The next enhancements to RETScreen Expert are currently under development in collaboration with Ontario's IESO and other key stakeholders. These software upgrades primarily focus on:

• Expansion of the depth and scale of the facility archetype database for municipalities, industries and university/college campuses;

• Integration of existing RETScreen 4 cogeneration and district energy models;

• Enhancement of data automation and connection capacities (RETScreen Connect) to better support the M&V of implemented projects;

• Improvements to the software to meet the ISO 50001 Energy Management Standard and the U.S.-DOE Superior Energy Performance (SEP) Standard.

Beyond the medium term, RETScreen Expert will eventually incorporate advanced artificial intelligence to further reduce the costs of assessing energy options.

The RETScreen software can be downloaded and RETScreen International contacted at www.retscreen.net.

Chapter 21

The Atmospheric Fund

Ms. Julia LANGER, The Atmospheric Fund (TAF)

A MUNICIPAL CLIMATE PIONEER

Climate action comes in many forms and must be implemented by many different players. One indispensable tool for reducing the world's carbon emissions is EE. It is equally clear that cities—where 70% of global emissions originate—must assume a leadership role. That is why city-level climate champions like The Atmospheric Fund (TAF) are crucial in advancing the policies, technologies and capital flows needed to accelerate EE.

TAF's story starts in 1988, before climate change was headline news, at a conference held in Toronto titled The Changing Atmosphere: Implications for Global Security. This prestigious gathering inspired several city councilors to figure out what to do locally about this global problem.

In 1991, an idea came to fruition: Toronto City Council voted to use the proceeds from the sale of a shuttered jail farm to create a $23 million endowment to tackle climate change and reduce air pollution. Established by provincial legislation, the Toronto Atmospheric Fund thus became the first municipal climate change agency in the world. Thanks to its unique structure, TAF has operated sustainably as a non-profit corporation at arm's length of the City without receiving any tax dollars.

EARLY INVESTMENTS

From the start, TAF focused on helping the City of Toronto lower its carbon emissions through EE:

- TAF's first financing project was a $15 million loan to the City of Toronto to upgrade street and lane lighting, which reduced carbon emissions by 20,000 tons and saved the City $2.7 million each year.

- In 1995, TAF lent Enwave District Energy $150,000 as seed money to develop the deep lake water cooling concept which now air conditions over 100 downtown buildings.

- In 2003, TAF structured a non-debt financing plan for the replacement of aging appliances with energy efficient units in close to 60,000 Toronto Community Housing units. Ultimately funded by the Federation of Canadian Municipalities and RBC and repaid through savings, this measure reduced Toronto's emissions by 5,500 tons per year.

Other investments in renewable energy and transportation initiatives like the wind turbine at Exhibition Place, the regional Smart Commute program, hybrid electric vehicle batteries, and a local car sharing service (now part of Enterprise) have paid off financially and environmentally. TAF was also a key funder of the campaign to phase out Ontario's coal-fired power generation, a step that dramatically reduced the carbon emissions of Toronto and Ontario.

APPROACH

There is significant return on investment from increasing EE, but what are the barriers? Often, a lack of familiarity with solutions like innovative heating and cooling technology constitutes a barrier, as well as the lack of access to capital and, ultimately, outdated policies that hold back the uptake of EE measures.

Over the years, TAF has utilized tools and expertise to help overcome these barriers through a three-pronged approach:

- Demonstrate and de-risk the feasibility of innovative technologies, policies and/or financing structures;

- Mobilize capital for low-carbon solutions by using TAF funds first;

- Spearhead policy and program improvements to scale up EE gains.

PRIORITY FOCUS

Buildings are responsible for 53% of Toronto's carbon emissions through electricity, natural gas and water consumption. With half of the Toronto population living in multi-unit residential buildings, improving high-rise EE has been a priority in TAF activities. As with other impact investments, TAF seeks not only a financial return, but also emission reductions and the successful demonstration and de-risking of climate innovations.

Existing Buildings

Retrofits can deliver financial and energy savings, but access to capital and so-called performance anxiety are common challenges. That is why TAF created the Energy Savings Performance Agreement (ESPA)—a shared financial savings agreement that allows building owners to improve their facilities and reduce energy consumption without borrowing or dipping into reserve funds. To scale up this innovative model, TAF incubated a separate for-profit entity—Efficiency Capital.

In tandem with financing, TAF demonstrates and de-risks key technologies through its flagship program TowerWise. Starting in 2009, TAF partnered with Toronto Community Housing and various coops to plan, implement and monitor the multi-measure retrofits of ten buildings. Through the installation of high-efficiency HVAC, lighting, water fixtures, thermostats, and other measures, the project improved indoor air quality, reduced energy waste and lowered operating costs, thereby yielding $646,000 in annual utility cost savings. Through such in-field projects, TAF and partners have demonstrated that EE gains of 20 to 30% are feasible when owners, engineers and investors strategically align. TAF will continue its collaboration with Toronto Community Housing for a second project phase that aims to generate even greater efficiency gains.

New Buildings

With an ongoing building boom in Toronto, especially for condominiums, it is critical to avoid locking in carbon emissions by fostering energy efficient construction. To this end, TAF partnered with the developer Tridel to co-create the Green Condo Loan, a financing tool that allows condo developers to incorporate energy efficient measures that exceed Ontario Building Code requirements without affecting the market price.

Demonstrating that it is possible to build much more energy efficient condos paved the way for TAF to advocate for stronger efficiency requirements for all new constructions. The Toronto Green Standard (TGS), approved by Toronto City Council in 2010, avoids an estimated 180,000 tons of carbon emissions per year. TGS in turn leveraged changes to the Ontario Building Code (OBC) which was modified to include Toronto's energy standards in 2012. In 2014, TGS 2.0 further strengthened requirements as the result of TAF's work. TAF is now working with the City and Province to entrench even more efficiency requirements in both TGS and OBC with a view to net zero.

THE PATH AHEAD

After 25 years of focusing on Toronto, TAF has expanded its geographic scope to the Greater Toronto and Hamilton Area (GTHA) with a new $17 million endowment from the Government of Ontario. TAF is now forging new ties with municipal governments, non-profits, building owners and other stakeholders to advance EE across Canada's fastest-growing urban region.

Simultaneously, TAF is deepening its partnership with the City of Toronto to reach the city's 2050 climate target (reduce GHG emissions by 80% by 2050). While TAF has helped decouple GHG emissions and economic growth, as well as lower carbon emissions by 23% since 1990, even as Toronto's population soared by 19%, more ambitious action is needed for Toronto to reach 80%. Dubbed TransformTO, this Council-approved strategy encompasses more than 30 low-carbon actions many of which prominently include EE retrofits and improved building standards. The scale of retrofits envisioned alone will create 80,000 person years of employment. With its emphasis on EE, the TransformTO plan is in line with international climate plans: reaching the Paris Agreement goals is only possible through EE since it constitutes the biggest, fastest, and cheapest low-carbon solution.

As TAF expands throughout the region, it will continue focusing on EE through a wide range of approaches: innovative financing for retrofits, improved construction standards, better planning and land use management, as well as building capacity to implement and monitor local measures. Through these efforts, TAF plays an important role in helping all levels of government reach their climate targets.

Conclusion

The Canadian EE sector has been evolving tremendously since its inception in the mid-1970s. Each province perceived the benefits of using EE as a resource to meet energy demand, but the sector evolved differently in each part of the country. Now that the fight against climate change has become one of the main drivers of the sector, things are changing rapidly and the national landscape is becoming more complex, as attested to by the research herein on each province, territory, as well as federal-sector policies, regulations and initiatives.

No one can gain comprehensive understanding of EE in Canada if no in-depth research is conducted on how each province develops institutional frameworks and policies. Hence the need for this book. Indeed since the energy sector as a whole is a provincial jurisdiction, the approaches adopted in each province and territory differ. Not only does progress in the use of EE differ greatly, but the legal and institutional framework is also very specific to each jurisdiction. This makes an analysis of the EE sector in Canada very difficult and a publication such as the Canadian Energy Efficiency Outlook so interesting.

As we analyzed the entire sector at the national, provincial and territorial levels, the diversity in approaches over time to stimulate the market toward increased EE became apparent. From the internal government structures (dedicated agencies, different ministries), to the use of utilities as delivery agents, to the roles of regulators, the Canadian EE sector significantly varies from coast to coast.

Nonetheless, the common denominator is the predominant use of grants and subsidies as a market transformation tool over the years. Other support mechanisms such as capacity building, adapted financing mechanisms and carbon trading schemes have not yet been used to their full potential and much remains untapped.

Other conclusions drawn while researching this book relate to the important changes made to the approaches used to foster the EE market in each jurisdiction due to political shifts over the years in favor of reducing the Canadian carbon footprint by increasing the importance of EE in the political landscape. Changes in policies and program offerings, as well as lack of continuity in structures, often contribute to the lack of sustainable success expected from market transformation initiatives. Specifically, the role of utilities ranges from playing a leading role in transforming the EE market to no role at all and everything in between depending on the jurisdiction, thereby sending conflicting signals to the market. It is also common for utilities and provincial agencies to attempt to achieve similar results using different tools, thus sometimes rendering the market offering too complex for energy end users to understand.

Despite all these challenges created by local circumstances, EE has gained a lot of ground in most parts of Canada over the years. It is now perceived as an essential factor in meeting the future energy demand of Canadians. EE is also an essential tool and a pillar of the Canadian strategy to fighting climate change.

There is a lot to learn from the last 40 years of Canada's experience in developing best EE practices. Lessons learned in one corner of the country can be transposed to another jurisdiction. The Canadian market is still fragmented because of the provincial nature of the country, and a lot would be gained by increasing exchanges and experiences between provinces and territories. It is clear that federal initiatives can be used as catalysts for greater shared interests and innovations.

Researching this book also bears out the important role that the EE community plays in supporting government initiatives. The chapters herein allow readers to discover many innovative private and non-profit organizations that are trying to break through barriers preventing the use of mechanisms such as EPC, EE financing, and capacity building among others. These mechanisms have significantly contributed to the development of the EE sector, and many more will likely be developed since governments encourage and help the EE market to achieve their sustainability and conservation objectives.

It has been a long journey to attain the current level of EE throughout Canada, and countless actors should be recognized for playing such important roles in this accomplishment. Yet, the journey is just beginning and much potential for more significant gains remains. Governments at all levels are multiplying their efforts to achieve more and more ambitious goals. Climate change-related initiatives will also greatly contribute to making the Canadian economy more energy efficient for many years to come. Moreover, non-profit organizations and the private sector will have to play a very important role in implementing projects and initiatives at all levels.

The future of EE in Canada has never been so bright and the sector will continue to evolve, but at a faster rate than ever before. The Canadian Energy Efficiency Outlook will hopefully contribute to this by helping all decision makers and stakeholders understand past efforts, current policies and initiatives in order to learn from those experiences.

Appendices

Authors

Ms. Geneviève GAUTHIER
Econoler
ggauthier@econoler.com
Geneviève Gauthier is an EE and RE expert with significant experience advising public and private organizations in the commercial, institutional and industrial sectors across Canada. She cumulates more than 17 years of experience in EE, RE and natural gas projects across Canada.

Ms. Gauthier advises clients on a great variety of policies, as well as the development and implementation of EE projects under the EPC approach in both the public and private sectors. She also collaborates on national projects for the development of regulations on mandatory energy audits, mandatory annual reporting of energy usage and action plans, minimum energy performance standards, EE labeling, measurement and verification, as well as process optimization monitoring and targeting. She has acted as lead in conducting numerous strategic studies such as EE financing mechanism design, demand-side potential studies, EE master plans, as well as small and large-scale feasibility studies.

She sits on the board of the CEEA and is an active member of several sectoral not-for-profit organizations involved in the EE market in Canada.

Ms. Gauthier is a Chemical Engineer who matriculated at École Polytechnique de Montréal and holds a Master's Degree in Paper Science and Engineering from the Institute of Paper Science and Technology (now merged with the Georgia Institute of Technology in Atlanta, Georgia). She is also certified by the Association of Energy Engineers (AEE) as a Certified Measurement and Verification Professional (CMVP), Certified Energy Auditor (CEA) and Certified Energy Manager (CEM). Her vast skills and knowledge have been recognized by many awards including the Jasper Mardon Memorial Prize (2011) and the Outstanding Master Student of the Year from the Institute of Paper Science and Technology at Georgia Institute of Technology (2004).

Mr. Pierre LANGLOIS
Econoler
planglois@econoler.com

Pierre Langlois is an internationally recognized EE expert who has cumulated 30 years of experience in the clean energy sector, specifically in the design, financing, implementation and monitoring of EE and RE projects and programs, both in Canada and abroad.

He is a first-rate Canadian and international expert in all EE subsectors, with impressive accomplishments in policy development, demand-side management, financing mechanisms, EPC and the design of innovative initiatives. He developed his expertise by working for national and international institutions, development agencies, governments, utilities, as well as international development banks. He gained unique experience in over 60 countries, having been a leading force in the use of EE at a time when the concept was not recognized as it is today.

Over the course of his career, Mr. Langlois has acted as member of the board of directors of several ESCOs throughout the world and a number of international organizations such as the Efficiency Valuation Organization (EVO), for which he has served on the board since 2005 and currently acts as chair.

His professional achievements have been awarded numerous prizes such as the Clean Technology Award 2000 jointly by the Climate Technology Initiative and the Organisation for Economic Co-operation and Development (OECD) at COP 6 (2000), and the Legend in Energy honor awarded by the Association of Energy Engineers (2007).

Mr. Langlois is a Mechanical Engineer who graduated from Laval University in 1986. He is a sought-after speaker for national and international events and has delivered over 250 conferences in Canada and around the world on different topics related to the EE sector. He is also a prolific author who has released several national and international publications including, among others, the following:

- The Bulgarian Energy Efficiency Fund: A Success Story and an Inspiring Example of Energy Efficiency Financing, 2017.

- State of Energy Report—Dubai 2014, Co-author of the Demand-Side Management chapter, Supreme Council of Energy, launched for Expo 2020 Dubai UAE, 2013.

- Joint Public-Private Approaches for Energy Efficiency Finance: Policies to Scale Up Private Sector Investment, International Energy Agency, 2012.

- Guidance on EE Financing in Public Buildings, European Investment Bank, 2012.

- ESCO World Outlook, Co-author, The Fairmont Press Inc., 2012.

- Seven best practice booklets for energy efficiency practitioners in the LAC Region, Inter-American Development Bank, 2012.

- ESCOs Around the World, Lessons Learned in 49 Countries, co-author, The Fairmount Press Inc., 2009.

- L'apport du partenariat public-privé dans le financement des projets en efficacité énergétique, L'institut Francophone du Développement Durable (IFDD) and the Agence de l'efficacité énergétique, 2009.

- Partenariat public-privé dans le financement de l'efficacité énergétique et du développement des énergies propres en Afrique, IEPF, 2006.

- Demand-side Management from a Sustainable Development Perspective—Experiences from Canada and India, IREDA, 2003.

- Guide to Working with Energy Service Companies in Central Europe, OECD and Climate Technology Initiative, 2003.

- The Evolution of ESCOs in Developing Countries and Economies in Transition, Energy Efficiency in Russia, Transfer of Canadian know-how Newsletter, fall 2002.

- ESCOs in Emerging Markets, the Bulletin on Energy Efficiency, IREDA, India, February 2001.

- Energy Efficiency Projects and their Financing Mechanisms, IEPF, 1999 (publication translated in Arabic by the United Nations, 2000).

- Energy Efficiency in Aluminum Plants Guidebook, Association de l'industrie de l'aluminium du Québec, 1998.

Contributing Authors

Mr. Andrew PAPE-SALMON (British Columbia)
Building and Safety Standards Branch, Government of British Colombia

Andrew Pape-Salmon is the Executive Director of the Building and Safety Standards Branch of the Government of British Columbia. He specializes in clean energy policy, economics, and systems design. Over the past 17 years, Mr. Pape-Salmon has worked extensively with the construction and real estate industries, energy utilities, and all levels of government to advance EE and climate leadership strategies for communities, buildings, and equipment. Between 2013 and 2017, he was an Associate and Senior Energy and Sustainability Specialist with RDH Building Science Inc.

Mr. Pape-Salmon is also a Professional Engineer with a Bachelor's Degree in Applied Science in Systems Design Engineering from the University of Waterloo and a Master's Degree in Resource and Environmental Management from Simon Fraser University. In 2016, he was inducted as a Fellow into the Canadian Academy of Engineering.

Ms. Brenda WALLACE (Saskatchewan)
City of Saskatoon

Ms. Wallace is Director, Environmental & Corporate Initiatives for the City of Saskatoon. She leads a team of environment professionals, engineers, and project managers responsible for initiatives that address the City of Saskatoon's Strategic Goal of Environmental Leadership and major city-building projects that enhance quality of life. There are four components to her environmental mandate: waste diversion; EE; green energy generation; and environmental protection.

Prior to joining the City of Saskatoon in 2010, Ms. Wallace was Resource Planning Manager at Meewasin Valley Authority for three years. From 1996 to 2007, she contracted out her professional services to a number of non-profit organizations. She led the Saskatchewan Economic Developers Association (SEDA) as Executive Director from 1996 to 2002 and was also the founding Executive Director for the Saskatoon Housing Initiatives Partnership (SHIP) from 1999 to 2007.

Ms. Wallace studied at the University of Saskatchewan in Regional and Urban Planning and at the School of Resource and Environmental Planning, Massey University, New Zealand.

Mr. Dany ROBIDOUX (Manitoba)
Eco-West Canada

Dany Robidoux is Executive Director of Eco-West Canada, an organization that aims to involve rural communities in greening their local economies by establishing GHG emission inventories and developing local climate change action plans, as well as supporting municipalities seeking green technology commercialization opportunities, attracting direct foreign investments, and training skilled workforces. In his position, Mr. Robidoux plays a key role in developing innovative cogeneration and biomass heating initiatives.

Mr. Robidoux holds both a B.A. and an Honours B.Comm. from the University of Manitoba. He is a member of CMC-Canada and is the Chair of the Manitoba Association of Regional Recyclers (MARR).

Mr. Dinesh S. PARAKH (RETScreen International)
Natural Resources Canada

Dinesh S. Parakh currently manages strategic partnerships, communications, business development, and training & capacity building for RETScreen International at Natural Resources Canada's CanmetENERGY research centre in Varennes, Quebec. He is also the lead author of the RETScreen Clean Energy Legal Toolkit and the RETScreen Clean Energy Policy Toolkit.

A lawyer in Ontario, Dinesh has worked in international development since 1999 with extensive experience throughout Africa. He was formerly with the United Nations Development Programme and World Food Programme in South Africa, as well as the Canadian International Development Agency (CIDA) in Ottawa, Canada. Dinesh holds a B.A. from Yale University, and an M.A. and J.D. from the University of Toronto.

Mr. Gregory LENG (RETScreen International)
Natural Resources Canada

Gregory J. Leng is the creator of the RETScreen Clean Energy Management Software and Director at Natural Resources Canada's CanmetENERGY research centre in Varennes, Quebec. Mr. Leng specializes in the clean energy market, technology and finance interface, and has been working in the renewable energy and EE sector since 1987.

Prior to joining NRCan, Mr. Leng was based in Hyderabad, India as the India Country Manager for the International Fund for Renewable Energy and Energy Efficiency. He obtained a Master of Science Degree from the University of Massachusetts Lowell in Energy Engineering (Solar Energy Engineering) and a Bachelor of Commerce Degree (Marketing and International Business) from McGill University.

Independent Electricity System Operator—IESO (Ontario)

The Independent Electricity System Operator (IESO) works at the heart of Ontario's power system, ensuring there is enough power to meet provincial energy needs in real time and managing the power system so that Ontarians have access to power when and where they need it. The IESO plans and prepares for future electricity needs and works with partners to guide conservation efforts under the direction of Terry Young, Vice-President of Policy, Engagement and Innovation.

Many conservation and EE experts contributed to the development of the Ontario chapter, yet the IESO wishes to acknowledge the contribution of Communications Advisor Jane Leckey who compiled the chapter on behalf of the IESO.

Mr. Jesse ROW (Alberta)
Alberta Energy Efficiency Alliance

Jesse Row is Vice President, Corporate Performance for the newly formed Energy Efficiency Alberta (EEA). Prior to joining EEA, he acted for nearly 10 years as the Executive Director of the Alberta Energy Efficiency Alliance (AEEA), a group of companies, municipalities, non-profit organizations and individuals working together to increase EE. In 2006, he helped start up the AEEA and establish it as a leading voice in the province on EE. His current work at the AEEA involves connecting those working on EE in the province and advocating for the creation of new EE policies, programs and practices in government and industry.

Mr. Row holds a B.Sc. in Mechanical Engineering from the University of Alberta and is a registered professional engineer in the Province of Alberta. He began his career working with the Pembina Institute on a number of consulting projects focused on economic, environmental and social life-cycle assessments of energy technologies and projects including fuel cell vehicles and biomass power plants.

Ms. Julia LANGER (TAF)
The Atmospheric Fund (TAF)

Julia Langer is the CEO of TAF, an agency dedicated to advancing solutions to climate change in the Greater Toronto and Hamilton Area. Under her leadership, TAF has accomplished a range of high-profile EE advancements. These include strengthening Toronto's and Ontario's building codes, demonstrating the feasibility of multi-measure retrofit technologies, financing via the TowerWise program, establishing energy and water reporting as well as benchmarking requirements, and securing the approval of TransformTO, Toronto's 2050 climate plan which prominently features EE actions.

Prior to joining TAF in 2009, Julia worked at World Wildlife Fund (WWF) Canada for 17 years. During that time, she established and led campaigns to address climate change, protect marine life, ban toxic chemicals, and advance sustainable agriculture.

In addition to her roles at TAF and WWF, Julia worked as an environmental analyst and campaigner for the following organizations: Pollution Probe; the Canadian chapter of Friends of the Earth; and the Federation of Canadian Municipalities. She also worked as a policy advisor to the Ontario Minister of the Environment.

Julia holds a Bachelor's Degree in Environmental Sciences from the University of Toronto.

Ms. Krista LANGTHORNE (Newfoundland and Labrador)
Newfoundland Power

Currently serving as Newfoundland Power's Manager of Energy Conservation, Krista Langthorne's responsibilities include planning activities to advance conservation at Newfoundland Power and overseeing the evaluation of customer energy conservation program offerings to identify areas of improvement. She also supervises the energy conservation team, including the marketing, planning, organization, development and delivery of residential and commercial programs and all customer outreach activities.

Krista Langthorne holds a Bachelor of Arts from Memorial University, an Advanced Diploma in Sustainable Energy and Building Technologies from Humber College, Associates Certificate in Business Analysis from Royal Roads University and Project Management Certificate from the University of Toronto.

Ms. Marie COUTURE-ROY (Evaluation)
Econoler
mcouture@econoler.com

Ms. Couture-Roy is Program Evaluation Director at Econoler. She has extensive experience in the EE sector and more specifically in the evaluation of energy management programs both in Canada and abroad. She has participated in the evaluation of an impressive number of EE programs in most provinces around Canada, as well as on the international market (Mexico and Chile), to measure energy impacts and help improve processes, among other tasks.

She holds a Bachelor's Degree in Mechanical Engineering from Laval University and is a frequent speaker at EE evaluation events in North America. She is also a published author of many papers on program evaluation at the national and international level.

Mr. Martin LUYMES (CEEA)
Heating, Refrigeration and Air Conditioning Institute of Canada (HRAI)

Martin Luymes is Director of Programs and Relations for the Heating, Refrigeration and Air Conditioning Institute of Canada (HRAI). He is responsible for the Association's government relations and industry advocacy work and oversees programs and services for the three membership divisions—manufacturers, wholesalers and contractors—as well as the HRAI Technical Training Division. Mr. Luymes is also Chair of the Canadian Energy Efficiency Alliance (CEEA), a Canadian independent EE advocate organization that works with governments and stakeholders to ensure EE is a priority for all sectors of the economy.

Mr. Mathieu CÔTE (Training)
Canadian Institute of Energy Training (CIET)
mcote@cietcanada.com

Mathieu Côte is the Canadian Institute of Energy Training (CIET) Executive Director, the most important dedicated EE training organization in the country. Mr. Côte has acquired significant knowledge in capacity building initiatives, from needs assessment to training curriculum development to certification processes. His experience in supporting the development of EE markets through training and capacity building initiatives covers all provinces in Canada as well as many countries around the world.

Mr. Côte holds an MBA in International Business. He has been a certified Business Energy Professional (BEP) since 2014 and a certified Project Management Professional (PMP) since 2012. He also served as a board member for the Canadian Energy Efficiency Alliance (CEEA) from 2014 to 2016.

Mr. Peter LOVE (EPC)
Energy Services Association of Canada
peter@energyservicesassociation.ca

Peter Love is President of the Energy Services Association of Canada, a leading advocate of performance-based solutions which represents energy service companies in Canada. He also serves on several corporate and non-profit advisory boards and boards of directors for the Ontario Climate Consortium (chair), Manitoba Race to Reduce, the Fortis BC Energy Efficiency and Conservation Advisory Group, as well as the newly formed Energy Efficiency Alberta. He also currently acts as Adjunct Professor at the York University Faculty of Environmental Studies where he teaches courses on energy and environmental policy.

Mr. Love received his BA and MBA degrees from the University of Toronto. He also holds the ICD.D designation from the Institute for Corporate Directors. Mr. Love has spent more than 40 years as a leader in Canadian energy policy. Among other roles, he has served as Ontario's first Chief Energy Conservation Officer for the Ontario Power Authority and as Executive Director of the Canadian Energy Efficiency Alliance.

Ms. Stephanie WHITEHEAD (Yukon)
Yukon Energy Corporation

Stephanie Whitehead is Energy Conservation Advisor at the Yukon Energy Corporation. She is in charge of developing and administering the inCharge DSM program, among other things. Ms. Whitehead is also part of Yukon Energy Resource Planning team and worked on the 2016 Resource Plan.

Prior to joining Yukon Energy in 2011, she worked as an environmental engineer in hydrology and environmental assessment in Alberta and the Yukon.

Ms. Whitehead studied at McGill University in Civil and Environmental Engineering.

Mr. Stephen MACDONALD (Nova Scotia)
EfficiencyOne

Stephen MacDonald is the Chief Executive Officer of Efficiency-One, the not-for-profit operator of Canada's first electricity efficiency utility as the franchise holder of Efficiency Nova Scotia. Under Mr. MacDonald's leadership, EfficiencyOne has continued its track record of meeting energy savings targets under budget, with a focus on continuous improvement, innovation, and customer satisfaction. Nova Scotia's results and innovative model have been cited by the influential International Energy Agency and the recent Pan-Canadian Framework on Clean Growth and Climate Change.

As a Chartered Professional Accountant prior to his appointment as EfficiencyOne CEO, Mr. MacDonald served as EfficiencyOne's Chief Operating Officer and held senior leadership positions with a Nova Scotia Crown Corporation and Grant Thornton.

Mr. Tom BERKHOUT (British Columbia)
BC Ministry of Energy and Mines

Tom Berkhout is Senior Policy Advisor for Energy Efficiency with the BC Ministry of Energy and Mines. Prior to joining the ministry, he completed his Ph.D. in Resource Management and Environmental Studies at the University of British Columbia. His Ph.D. thesis focused on policies and practices to support transformative EE and conservation in the built environment.

Glossary of Acronyms

ACCO	Alberta Climate Change Office
ACEEE	American Council for an Energy Efficiency Economy
AEA	Arctic Energy Alliance
AEE	Agence de l'efficacité énergétique (QC)
AEE	Association of Energy Engineers
AEEA	Alberta Energy Efficiency Alliance
AETP	Alternative Energy Technologies Program
AGPI	Association des gestionnaires de parcs immobiliers (QC)
ASBA	Alberta School Boards Association
ASBOA	Association of School Business Officials of Alberta
AUC	Alberta Utilities Commission
BBB	Build Better Buildings Policy
BC	British Columbia
BCUC	British Columbia Utilities Commission
BD3	Boundary Dam Power Station Unit 3
BECC	Building Energy Codes Collaborative
BEE	Bureau de l'efficacité énergétique (QC)
BEIE	Bureau de l'efficacité et de l'innovation énergétiques (QC)
BNI	Business, Non-Profit and Institutional (Alberta Energy Savings Program)
BOMA	Building Owners and Managers Association of Canada
BPS	Broader public sector
CAA	Clean Air Agenda
CAC	Consumers' Association of Canada
CARF	Capital Asset Retrofit Fund
CBDR	Capacity-Based Demand Response program
CBERP	Community Building Energy Retrofit Program
CBSA	Canada Border Services Agency
CCA	Capital Cost Allowance
CCBFC	Canadian Commission on Building and Fire Codes
CCEMC	Climate Change and Emissions Management Corporation
CCEMF	Climate Change and Emissions Management Fund
CCNL	Conservation Corps of Newfoundland and Labrador
CCS	Carbon capture and sequestration
CDM	Conservation and demand management
CEA	Clean Energy Act
CECEP	Commercial Energy Conservation and Efficiency Program

CEEA   Canadian Energy Efficiency Alliance
CEF   Clean Economy Fund
CEM   Certified Energy Manager
CEMET   Canadian Energy Management and Environmental Training
          Program
CER   Community Expenditure Report
CFF   Conservation First Framework
CFIC   Conservation First Implementation Committee
CGS   Community and Government Services
CHBA   Canadian Home Builders' Association
CHP   Combined heat and power (cogeneration)
CIDA   Canadian International Development Agency
CIET   Canadian Institute for Energy Training
CII   Commercial, institutional and industrial
CIP   Capital Investment Plan
CIPEC   Canadian Industry Program for Energy Conservation
CLASP   Collaborative Labeling and Appliance Standards Program
CMF   Control Management Framework (NWT)
CMHC   Canada Mortgage and Housing Corporation
CMVP   Certified Measurement and Verification Professional
CNG   Compressed natural gas
COF   Conference of the Federation
COP   Conference of the Parties
CPR   Conservation potential review
CRA   Canada Revenue Agency
CREP   Community Renewable Energy Program
CSA   Canadian Standard Association
CTS   Cap-and-trade system
DEE   Direction de l'efficacité énergétique (Office of Energy Effi-
          ciency)
DR   Demand response
DSM   Demand-side management
EC   Efficiency Capital
ECCC   Environment and Climate Change Canada
ECHO   Environmental and Cultural Hiring Opportunity
ECM   Electrically commutated motor
ECO   Environmental Commissioner of Ontario
ECPA   Energy and Climate Partnership of the Americas
ECSTF   Electricity Conservation & Supply Task Force

EDA	Electricity Distributors Association
EE	Energy efficiency
EEA	Energy Efficiency Act
EEIP	Energy Efficiency Incentive Program
EERE	Office of Energy Efficiency and Renewable Energy
EESR	Energy Efficiency Standards Regulation
EFC	Electro-Federation Canada
EMMC	Energy and Mines Ministers' Conference
EM&V	Evaluation, measurement and verification
ENS	Efficiency Nova Scotia
ENSC	Efficiency Nova Scotia Corporation
EPC	Energy performance contracting/contract
EPP	Energy Performance Program
ESCO	Energy service company
ESPA	Energy Savings Performance Agreement
ESPC	Energy savings performance contract
ESPM	ENERGY STAR® Portfolio Manager
EUB	Energy and Utilities Board (New Brunswick)
EV	Electric vehicle
EVO	Efficiency Valuation Organization
EWRB	Energy and Water Reporting and Benchmarking
FBI	Federal Buildings Initiative
FCM	Federation of Canadian Municipalities
FSDS	Federal Sustainable Development Strategy
GEA	Green Energy Act
GEERS	Guelph Energy Efficiency Retrofit Strategy
GEF	Global Environment Facility
GGAP	Go Green Action Plan
GHG	Greenhouse gas
GIF	Green Investment Fund
GMF	Green Municipal Fund
GN	Government of Nunavut
GNP	Gross National Product
GNWT	Government of the Northwest Territories
GTF	Gas Tax Fund
GTHA	Greater Toronto and Hamilton Area
GWh	Gigawatt hours
HEELP	Home Energy Efficiency Loan Program
HELP	Home Energy Low-Income Program

HESP	Home Energy Savings Program
HRAI	Heating, Refrigeration and Air Conditioning Institute of Canada
HRV	Heat recovery ventilators
HVAC	Heating, ventilation and air conditioning
Hydro	Newfoundland and Labrador Hydro
IAP	Industrial Accelerator Program
ICE	Innovative Clean Energy Fund
ICI	Industrial Conservation Incentive
ICSP	Integrated Community Sustainability Plan
IEPEC	International Energy Program Evaluation Conference
IESO	Ontario's Independent Electricity System Operator
IIS	Island Interconnected System
IO	Infrastructure Ontario
IPEEC	International Partnership for Energy Efficiency Cooperation
IPMVP	International Performance Measurement and Verification Protocol
IPSP	Integrated Power System Plan
IRAC	Island Regulatory and Appeals Commission
IRP	Integrated resource plan
ktoe	kiloton of oil equivalent
kW	Kilowatts
kWh	Kilowatt hours
LDCs	Local distribution companies
LED	Light-emitting diode
LEED	Leadership in Energy and Environmental Design Program
LNG	Liquefied natural gas
LTEP	Long-term energy plan
LTRP	Long-term resource plan
LUEC	Levelized Unit Electricity Cost
M&V	Measurement and verification
MACA	Department of Municipal and Community Affairs
MBC	Manitoba Building Code 2011
MDM/R	Meter Data Management/Repository
MECB	Manitoba Energy Code for Buildings
MECC	Ministerial Energy Coordinating Committee
MEM	Ministry of Energy and Mines (British Columbia)
MEP	Municipal Energy Plan
MEPS	Minimum energy performance standards

MERN	Ministry of Energy and Natural Resources
MNECB	Model National Energy Code of Canada for Buildings
MOI	Ontario Ministry of Infrastructure
MOU	Memorandum of understanding
MT	Market transformation
MTRC	Modified Total Resource Cost
MUSH	Municipalities, Universities, Schools and Hospitals
MW	Megawatts
NAIMA	North American Insulation Manufacturers Association
NAPCC	National Action Program on Climate Change
NASA	National Aeronautics and Space Administration
NASEO	National Association of State Energy Officials
NBCC	National Building Code of Canada
NEB	Non-energy benefits
NEC	Nunavut Energy Centre
NECB	National Energy Building Code of Canada for Buildings
NEIA	Newfoundland and Labrador Environmental Industry Association
NERP	Nunavut Energy Retrofit Program
NEUD	National Energy Use Database
NHC	Nunavut Housing Corporation
NPV	Net Present Value
NRC	National Research Council Canada
NRCan	Natural Resources Canada
NTPC	Northwest Territories Power Corporation
NWT	Northwest Territories
OBC	Ontario Building Code
OCC	Office of Climate Change
OEB	Ontario Energy Board
OEE	Office of Energy Efficiency (NRCan)
OEE	Office of Energy Efficiency (PEI)
OERD	Office of Energy Research and Development
OGGO	Office of Greening Government Operations
OHBA	Ontario Home Builders' Association
OPA	Ontario Power Authority
OPO	Ontario Planning Outlook
OPS	Ontario Public Service
Pa	Pascals
PAC	Program Administrator Cost

PACE	Property-Assessed Clean Energy
PEI	Prince Edward Island
PERD	Program of Energy Research and Development
PPESCOs	Public purpose energy service companies
PSPC	Public Services and Procurement Canada
PSP	Power System Plan (NWT)
PST	Provincial Sales Tax
PUB	Public Utilities Board (MB)
PUB	Board of Commissioners of Public Utilities (NFL)
PV	Photovoltaic
PWS	Public Works and Services (NWT)
QEC	Qulliq Energy Corporation
RBC	Royal Bank of Canada
RE	Renewable energy
REEEP	Renewable Energy and Energy Efficiency Partnership
REEP	Residential Energy Efficiency Program (Newfoundland and Labrador)
RES	Renewable energy source
RFP	Request for proposals
RRSP	Registered Retirement Savings Plan
SCC	Standards Council of Canada
SCEE	Steering Committee on Energy Efficiency
SDGs	Sustainable Development Goals
SECDA	Saskatchewan Energy Conservation and Development Association
SEP	Superior Energy Performance
SES	Saskatchewan Environmental Society
SGER	Specified Gas Emitter Regulation
SHEU	Survey of Household Energy Use
SL&P	Saskatoon Light and Power
SOE	Save on Energy
StatCan	Statistics Canada
TAF	The Atmospheric Fund
TBS	Treasury Board Secretariat
TCHC	Toronto Community Housing Corporation
TÉQ	Transition Énergétique Québec (Quebec Energy Transition)
TGS	Toronto Green Standard
toe	Ton of oil equivalent
TRC	Total Resource Cost

TWh	Terawatt hours
UARB	Utility and Review Board
UCA	Utilities Commission Act
UMP	Uniform Methods Project
UN	United Nations
UNEP	United Nations Environment Programme
WCI	Western Climate Initiative
WDM	Western Development Museum

References by Chapter

Chapter 1

[1]   Statistics Canada. (2017). Supply and demand of primary and secondary energy in terajoules. (CANSIM 128-0016). Retrieved from http://www5.statcan.gc.ca/cansim/a26?lang=eng&id=1280016.

[2]   Industrial Efficiency Policy Database (IEPD). (2017). CA-2: Canadian Industry Programme for Energy Conservation (CIPEC). Retrieved from http://iepd.iipnetwork.org/policy/canadian-industry-programme-energy-conservation-cipec.

[3]   Collaborative Labeling and Appliance Standards Program (CLASP), (2005). Energy-Efficiency Labels and Standards. A Guidebook for Appliances, Equipment, and Lighting. 2nd Edition Retrieved from http://pdf.usaid.gov/pdf_docs/Pnadj148.pdf

[4]   Wiel, S., McMahon, J.E. (2005). Energy-Efficiency Labels and Standards. A Guidebook for Appliances, Equipment, and Lighting (2nd Edition). Washington, D.C.: CLASP. Retrieved from https://escholarship.org/content/qt01d3r8jg/qt01d3r8jg.pdf.

[5]   NRCan. (2016). Improving Energy Performance in Canada: Report to Parliament under the Energy Efficiency Act. 2013–2015. Retrieved from http://oee.nrcan.gc.ca/publications/statistics/parliament/2013-2015/pdf/parliament13-15.pdf.

[6]   Energy Efficiency Regulations. (2016). Canada Gazette, Part I, April 30, 2016, Vol. 140, No. 18. Retrieved from http://www.gazette.gc.ca/rp-pr/p1/2016/2016-04-30/html/reg1-eng.php.

[7]   Canada Mortgage and Housing Corporation. (2017). CMHC Green Home: Helping to Make Energy-Efficient Housing Choices More Affordable. Retrieved from https://www.cmhc-schl.gc.ca/en/hoficlincl/moloin/hopr/hopr_003.cfm.

[8]   Environment and Climate Change Canada (ECCC), 2016. A Federal Sustainable Development Strategy for Canada. 2016-2019

[9]   NRCan. (2016). Energy efficiency—buildings. Retrieved from http://www.nrc-cnrc.gc.ca/eng/solutions/advisory/codes_centre/faq/energy_necb.html.

[10] NRCan. (2013). Technical Guide to Class 43.1 and 43.2 (2013 Ed.). Retrieved from https://www.nrcan.gc.ca/sites/www.nrcan.gc.ca/files/energy/pdf/Class_431-432_Technical_Guide_en.pdf.

[11] NRCan. (2013). Energy Performance Contracting: Guide for Federal Buildings. Retrieved from https://www.nrcan.gc.ca/

sites/www.nrcan.gc.ca/files/oee/files/pdf/communities-government/buildings/federal/pdf/12-0419%20-%20EPC_e.pdf.

Chapter 2

[1] Energy Efficiency Branch of Alberta Energy, An Assessment of Energy Efficiency in Alberta (1993): pp. 61-65.
[2] Market Wired. (2014). Alberta's Climate Change Central to Close. Retrieved from http://www.marketwired.com/press-release/albertas-climate-change-central-to-close-1916234.htm.
[3] City of Medicine Hat. (s.d.). HAT Smart. Retrieved from https://www.medicinehat.ca/government/departments/utility-sustainability/hat-smart.
[4] Alberta Infrastructure. (n.d.). Environmental Certifications. Retrieved from http://www.infrastructure.alberta.ca/3814.htm.
[5] Alberta Environment and Parks. (2008). Alberta's 2008 Climate Change Strategy: Responsibility / Leadership / Action. Retrieved from http://aep.alberta.ca/forms-maps-services/publications/documents/AlbertaClimateChangeStrategy-2008.pdf.
[6] Alberta Government. (2017). Specified Gas Emitters and Reporting 2016 Workshop. Retrieved from http://aep.alberta.ca/climate-change/guidelines-legislation/specified-gas-emitters-regulation/documents/2016-SGER-WorkshopPresentation-Mar24-2017.pdf.
[7] Alberta Government. (2017). City Charters: Overview Package. Retrieved from https://www.alberta.ca/documents/City-Charters-Overview-Package.pdf.

Chapter 3

[1] Statistics Canada. 2017. Report on Energy and Supply Demand in Canada: 2015 Preliminary. Catalog no. 57-003-X. Retrieved from http://www.statcan.gc.ca/pub/57-003-x/57-003-x2017002-eng.htm.
[2] Supra note 1.
[3] Blumstein, C., Goldstone, S., & Luztenhiser, L. 2000. A theory-based approach to market transformation energy. Energy Policy, 28(2), 137-144. https://doi.org/10.1016/S0301-4215(99)00093-2.
[4] British Columbia Government (2011). Review of BC Hydro. Retrieved from https://news.gov.bc.ca/files/Newsroom/downloads/bchydroreview.pdf.

[5]   FortisBC Energy Utilities. 2014. 2014 Long Term Resource Plan. P.
      23 of EEC Plan. Appendix I of LTRP. Start at page 1087 of 1444 in
      Volume 2.
[6]   FortisBC Inc. 2016. 2016 Long-Term Demand-Side Plan. Included
      in 2016 Long-Term Electric Resource Plan. Volume 2.
[7]   BC Hydro. 2016. Demand-Side Management Expenditures. Chap-
      ter 10 of the Fiscal 2017 to Fiscal 2019 Revenue Requirements Ap-
      plication.
[8]   FortisBC Energy Utilities. 2014. 2014 Long Term Resource Plan. P.
      23 of EEC Plan. Appendix I of LTRP. Start at page 1087 of 1444 in
      Volume 2.
[9]   FortisBC Inc. 2017. Electricity Demand-Side Management Pro-
      grams 2016 Annual Report. Retrieved from https://www.fortisbc.
      com/About/RegulatoryAffairs/GasUtility/NatGasBCUCSub-
      missions/Documents/170331_FBC_2016_DSM_Annual_Report_
      FF.pdf.
[10] BC Ministry of Energy and Mines. 2014. Guide to the Demand-Side
      Measures Regulation. Retrieved from https://www2.gov.bc.ca/
      assets/gov/farming-natural-resources-and-industry/electrici-
      ty-alternative-energy/energy-efficiency/guide_to_the_dsm_reg-
      ulation_july_2014_c2.pdf.
[11] Government of British Columbia. BC Energy Step Code. Retrieved
      from http://www2.gov.bc.ca/gov/content/industry/construc-
      tion-industry/building-codes-standards/energy-efficiency/ener-
      gy-step-code.
[12] BC Hydro. 2014. "Table 1 Expenditures and Incremental Electricity
      Savings for F2014." Report on Demand-Side Management Activi-
      ties for F2014. In Appendix Y (Attachment 1) of the Fiscal 2017 to
      Fiscal 2019 Revenue Requirements Application.
[13] BC Hydro. 2015. "Table 1 Expenditures and Incremental Electricity
      Savings for F2014." Report on Demand-Side Management Activi-
      ties for F2014. In Appendix Y (Attachment 2) of the Fiscal 2017 to
      Fiscal 2019 Revenue Requirements Application.
[14] For 2015-2016 to 2018-2019: BC Hydro. 2016. Table 1 "Cumulative
      Energy Savings at Customer Meter (GWh/yr)" in Appendix W—
      Demand-Side Management Data Tables of the Fiscal 2017 to Fiscal
      2019 Revenue Requirements Application.
[15] Supra note 12.
[16] BC Hydro. 2015. "Table 1 Expenditures and Incremental Electricity
      Savings for F2014." Report on Demand-Side Management Activi-

ties for F2014. In Appendix Y (Attachment 2) of the Fiscal 2017 to Fiscal 2019 Revenue Requirements Application.

[17]   For 2015-2016 to 2018-2019: BC Hydro. 2016. Table 5 "Total BC Hydro Costs ($ Million)" in Appendix W—Demand-Side Management Data Tables of the Fiscal 2017 to Fiscal 2019 Revenue Requirements Application.

[18]   The FortisBC Energy Utilities. 2015. Energy Efficiency and Conservation Program. 2014 Annual Report. Retrieved from https://www.fortisbc.com/About/RegulatoryAffairs/GasUtility/NatGasBCUCSubmissions/Documents/150330_FEU_2014_EEC_Annual_Report_FF.pdf.

[19]   FortisBC Energy Inc. 2016. Natural Gas Demand-Side Management Programs. 2015 Annual Report. Retrieved from https://www.fortisbc.com/About/RegulatoryAffairs/GasUtility/NatGasBCUCSubmissions/Documents/160330_FEI_2015_DSM_Annual_Report_FF.pdf.

[20]   FortisBC Energy Inc. 2017. Natural Gas Demand-Side Management Programs. 2016 Annual Report. Retrieved from https://www.fortisbc.com/About/RegulatoryAffairs/GasUtility/NatGasBCUCSubmissions/Documents/170331_FEI_2016_DSM_Annual_Report_FF.pdf.

[21]   FortisBC Energy Inc. 2013. Application for Approval of a Multi-Year Performance Based Ratemaking Plan for 2014 through 2018—Appendix I: Energy Efficiency and Conservation and Demand-Side Management. Retrieved from https://www.fortisbc.com/About/RegulatoryAffairs/ElecUtility/Documents/FBC_2014-2018_PBR_Plan_Volume_1-Application_REDACTED_FF.pdf.

Chapter 4

[1]   Manitoba Hydro. (n.d.). The History of Electric Power in Manitoba. Retrieved from https://www.hydro.mb.ca/corporate/history/history_of_electric_power_book.pdf.

[2]   Dunsky Energy Consulting. (2014). The Role and Value of Demand-Side Management in Manitoba Hydro's Resource Planning Process. Manitoba Public Utilities Board (Testimony of Philippe D). Retrieved from http://www.pubmanitoba.ca/v1/nfat/pdf/demand_side_management_dunsky.pdf.

[3]   Supra note 2.

[4]   Bob Brennan. (1991). Introducing Power Smart, our new approach

to conserving energy. Hydrogram Newsletter.

[5]   Manitoba Hydro Hydrogram Newsletter, April 24, 1991.

[6]   Manitoba Hydro. (2016). Helping Manitobans Move Toward A More Energy Sustainable Future: 2016/17 Demand-Side Management Plan. Retrieved from https://www.hydro.mb.ca/corporate/pdfs/demand_side_management_plan.pdf.

[7]   Ibid, p. 1.

Chapter 5

[1]   Statistics Canada. (2017). Supply and demand of primary and secondary energy in terajoules. (CANSIM 128-0016). Retrieved from http://www5.statcan.gc.ca/cansim/a26?lang=eng&id=1280016.

[2]   Statistics Canada. (2017). Report on Energy and Supply Demand in Canada: 2015 Preliminary. (Catalog no. 57-003-X). Retrieved from http://www.statcan.gc.ca/pub/57-003-x/57-003-x2017002-eng.htm.

[3]   Natural Resources Canada. (2016). New Brunswick's Electric Reliability Framework. Retrieved from http://www.nrcan.gc.ca/energy/electricity-infrastructure/18832.

[4]   New Brunswick Natural Resources and Energy. (n.d.). White Paper: New Brunswick Energy Policy. Retrieved from http://www2.gnb.ca/content/dam/gnb/Departments/en/pdf/Publications/NBEnergyPolicyWorkingGroup.pdf.

[5]   Energy Efficiency and Conservation Agency of New Brunswick (Efficiency NB). (2014). Annual Report 2013-2014. Retrieved from http://www2.gnb.ca/content/dam/gnb/Departments/en/pdf/Publications/EnergyEfficiency_ConservationAgency_Annual%20Report-e.pdf.

[6]   CBC News. (2014, December 12). Efficiency New Brunswick folded into NB Power. Retrieved from http://www.cbc.ca/news/canada/new-brunswick/efficiency-new-brunswick-folded-into-nb-power-1.2871099.

[7]   Department of Energy and Mines. (2016). Energy and Mines Annual Report: 2015-2016. Retrieved from http://www2.gnb.ca/content/dam/gnb/Departments/en/pdf/Publications/EnergyandMines_AnnualReport_2015-16.pdf.

[8]   New Brunswick Department of Energy. (2011). The New Brunswick Energy Blueprint. Retrieved from http://normandmousseau.com/documents/Locke-2.pdf.

[9]   New Brunswick Department of Energy and Mines. (2014). 2014/15-
      2016/17 Electricity Efficiency Plan. Retrieved from http://www2.
      gnb.ca/content/dam/gnb/Departments/en/pdf/Publications/
      EfficiencyPlanExecutiveSummary.pdf.
[10] Supra note 3.
[11] New Brunswick Energy & Utilities Board. (n.d.). New Brunswick
      Energy & Utilities Board. Retrieved from http://www.nbeub.ca/.
[12] Supra note 4.
[13] New Brunswick Department of Energy and Mines. (2014). The
      New Brunswick Energy Blueprint Final Progress Report. Retrieved
      from http://www2.gnb.ca/content/dam/gnb/Departments/en/
      pdf/Publications/EnergyBlueprintFinalProgressReport.pdf.
[14] University of New Brunswick (UNB). (n.d.). Energy Fundamentals
      for Leaders: Making informed decisions about energy issues. Re-
      trieved from http://www.unb.ca/saintjohn/sjcollege/profession-
      aldev/energy.html.
[15] New Brunswick Government. (n.d.) Transitioning to a Low-Car-
      bon Economy: New Brunswick's Climate Change Action Plan.
      Retrieved from http://www2.gnb.ca/content/dam/gnb/Depart-
      ments/env/pdf/Climate-Climatiques/TransitioningToALowCar-
      bonEconomy.pdf.
[16] Energy Efficiency Act: New Brunswick Regulation 95-70. (1995,
      O.C. 95-555). Retrieved from http://laws.gnb.ca/en/showfull-
      doc/cr/95-70//20171122.
[17] Supra note 9.
[18] NB Power. (n.d.). Energy Efficiency programs. Retrieved from
      https://www.nbpower.com/en/smart-habits/energy-efficien-
      cy-programs.
[19] Natural Resources Canada. (n.d.). Directory of Energy Efficien-
      cy and Alternative Energy programs in Canada. Retrieved from
      http://oee.nrcan.gc.ca/corporate/statistics/neud/dpa/poli-
      cy_e/results.cfm?programtypes= 4&regionaldeliveryid=5&at-
      tr=0.

Chapter 7
[1]   Government of Northwest Territories. NWT Bureau of Statistics.
      Retrieved from http://www.statsnwt.ca/.
[2]   Statistics Canada. (2017). Supply and demand of primary and sec-
      ondary energy in terajoules. (CANSIM 128-0016). Retrieved from

http://www5.statcan.gc.ca/cansim/a26?lang=eng&id=1280016.

[3]   Northwest Territories Power Corporation. (n.d.). Corporate Struc-
      ture. Retrieved from https://www.ntpc.com/about-ntpc/corpo-
      rate-structure.

[4]   Northwest Territories Government. (2011). Northwest Territories
      Energy Report. Retrieved from http://www.assembly.gov.nt.ca/
      sites/default/files/11-05-20td36-166.pdf.

[5]   Government of Canada. (2005). Canada—Northwest Territories
      Agreement on the Transfer of Federal Gas Tax Revenues for Cities
      and Communities 2005-2015. Control Management Framework,
      2009.

[6]   Northwest Territories Environment and Natural Resources. (2007).
      NWT Greenhouse Gas Strategy 2007-2011. Retrieved from http://
      www.enr.gov.nt.ca/sites/enr/files/strategies/greenhouse_gas_
      strategy_final.pdf.

[7]   Government of Northwest Territories Public Works and Services.
      (2016). Energy Conservation Initiatives Report: 2015-2016. Re-
      trieved from http://www.assembly.gov.nt.ca/sites/default/files/
      td_96-182.pdf.

[8]   AEA annual reports.

[9]   Supra note 4.

[10] Government of Northwest Territories. (n.d.). Retrieved from
      http://www.gov.nt.ca.

[11] Northwest Territories Government. (2010). Territorial Power Sub-
      sidy Policy. Retrieved from http://www.fin.gov.nt.ca/forms-doc-
      uments/territorial-power-subsidy-policy.

[12] Government of Northwest Territories Environment and Natural
      Resources. (n.d.). About Environment and Natural Resources.
      Retrieved from http://www.enr.gov.nt.ca/en/about-environ-
      ment-and-natural-resources.

[13] Government of Northwest Territories Infrastructure. (n.d.). Man-
      date and Responsibilities. Retrieved from https://www.inf.gov.
      nt.ca/en/mandate-and-responsibilities.

[14] Arctic Energy Alliance. (n.d.). About Us. Retrieved from http://
      aea.nt.ca/about-us.

[15] Government of the Northwest Territories. (2015). GNWT Response
      to the 2014 NWT Energy Charrette Report. Retrieved from http://
      www.assembly.gov.nt.ca/sites/default/files/td_271-175.pdf.

[16] Government of Northwest Territories. (2013). Northwest Territo-

ries Energy Action Plan. Retrieved from http://www.mrif.gouv.
qc.ca/PDF/actualites/nwt_energy_action_plan_december2013.
pdf.

[17] Governments of the Northwest Territories, Nunavut, and Yukon.
(2011). Paths to a Renewable North: A Pan-territorial renewable
energy inventory. Retrieved from http://www.anorthernvision.
ca/documents/RenewableEnergyInventoryEN.pdf.

[18] Paul Tukker. (2016, October 3). Carbon pricing scheme to reflect
North's 'specific challenges'. CBC News. Retrieved from http://
www.cbc.ca/news/canada/north/carbon-tax-northern-territo-
ries-reaction-1.3789586.

[19] Government of Northwest Territories. (2016). Energy Strategy Discussion Guide. Retrieved from https://www.inf.gov.nt.ca/sites/
inf/files/resources/11523-_energy_discusson_guide_web_.pdf.

[20] Ecology North. (2015). Carbon Pricing in the NWT. Retrieved from
http://ecologynorth.ca/wp-content/uploads/2016/04/CAR-
BON-PRICING-IN-THE-NWT-.pdf.

[21] Supra note 19.

[22] Arctic Energy Alliance. (n.d.). Retrieved from http://aea.nt.ca.

[23] Government of Northwest Territories. (2011). Good Building Practice for Northern Facilities. Third Edition.

[24] Supra note 4.

[25] Northwest Territories Environment and Natural Resources. (2011).
A Greenhouse Gas Strategy for the Northwest Territories 2011-
2015. Retrieved from http://www.enr.gov.nt.ca/sites/enr/files/
strategies/ghg_strategy_2011-2015.pdf.

[26] Ibid.

[27] Government of Northwest Territories. (n.d.). NWT Greenhouse
Gas (GHG) Emissions. Retrieved from http://www.enr.gov.nt.ca/
sites/enr/files/ccsf_ghg_emissions_factsheet.pdf.

[28] Ministerial Energy Coordinating Committee. (2008). Energy Priorities Framework. Retrieved from https://www.inf.gov.nt.ca/
sites/inf/files/energy_priorities_framework.pdf.

[29] Supra note 11.

[30] Natural Resources Canada. (2015). Codes: Northwest Territories.
Retrieved from http://www.nrcan.gc.ca/energy/efficiency/
buildings/capacity-building-resources/learn-more/codes/4235.

[31] Government of Northwest Territories. (2016.). Energy Strategy
Discussion Guide. Retrieved from http://www.pws.gov.nt.ca/

en / files / energy-strategy-discussion-guide.

[32] Arctic Energy Alliance. (2016). 2015-2016 Annual Report. Retrieved from http:/ / aea.nt.ca/blog/2016/07/annual-report-2015-2016.

[33] Supra note 15.

Chapter 8

[1] Statistics Canada. (2017). Supply and demand of primary and secondary energy in terajoules. (CANSIM 128-0016). Retrieved from http:/ /www5.statcan.gc.ca/cansim/a26?lang=eng&id=1280016.

[2] Efficiency Nova Scotia Corporation Act (SNS 2009, c 3). Retrieved from https:/ /nslegislature.ca/sites/default/files/legc/statutes/efficiency%20ns%20corporation.pdf.

[3] Public Utilities Act (RSNS 1989, c 380 s 79). Retrieved from https:/ /nslegislature.ca/sites/default/files/legc/statutes/public%20utilities.pdf.

[4] David Wheeler. (2008). Stakeholder Consultation Process for an Administrative Model for DSM Delivery in Nova Scotia. Halifax, NS: Dalhousie University. Retrieved from http:/ /0-nsleg-edeposit.gov.ns.ca.legcat.gov.ns.ca/deposit/b10579424.pdf.

[5] Electricity Efficiency and Conservation Restructuring Act (SNS 2014, c 5). Retrieved from https:/ /nslegislature.ca/sites/default/files/legc/statutes/electricity%20efficiency%20and%20conservation%20restructuring%20(2014).pdf.

[6] Supra note 3 at s 79C(2)(a).

[7] Ibid at s 79Q(1-2).

[8] Ibid at s 79Q(2).

[9] Ibid at s 79R(4).

[10] Supra note 2.

[11] NS Power. (2017). How we make electricity. Retrieved from http:/ /www.nspower.ca/en/home/about-us/how-we-make-electricity/default.aspx.

[12] Ibid.

[13] Ibid.

[14] Nova Scotia Department of Energy. (2014). Using Less Energy: Nova Scotia's Electricity Efficiency and Conservation Plan. Retrieved from http:/ /0-nsleg-edeposit.gov.ns.ca.legcat.gov.ns.ca/deposit/b10670427.pdf.

[15] EfficiencyOne. (2017). Unleash the power of efficiency: Strategic Plan 2016-2020. Retrieved from https:/ /www.efficiencyone.ca/

wp-content/uploads/2017/12/EfficiencyOne-2016-2020-Strategic-Plan.pdf

[16] Ibid.

[17] Climate Change Nova Scotia. (n.d.). Nova Scotia Cap and Trade Program Design Options. Retrieved from https://climatechange.novascotia.ca/sites/default/files/Cap-and-Trade-Document.pdf.

[18] Environmental Goals and Sustainable Prosperity Act. (SNS 2007, c 7). Retrieved from https://nslegislature.ca/sites/default/files/legc/statutes/environmental%20goals%20and%20sustainable%20prosperity.pdf.

[19] Nova Scotia Environment. (2009). Toward a Greener Future: Nova Scotia's Climate Change Action Plan. Retrieved from https://climatechange.novascotia.ca/sites/default/files/uploads/ccap.pdf.

[20] Regulations Respecting Greenhouse Gas Emissions (NS Reg. 260/2009). Retrieved from https://www.canlii.org/en/ns/laws/regu/ns-reg-260-2009/latest/ns-reg-260-2009.html.

[21] Nova Scotia Department of Energy. (2010). Renewable Electricity Plan: A path to good jobs, stable prices, and a cleaner environment. (At 2; Renewable Electricity Regulations, NS Reg 155/2010, s 6-6A). Retrieved from http://normandmousseau.com/documents/Locke-3.pdf.

[22] Efficiency Nova Scotia. (n.d.). Efficiency Trade Network General Participation Agreement. Retrieved from https://www.efficiencyns.ca/wp-content/plugins/effone-etn/etn-app/dist/assets/general-participation-agreement.pdf.

[23] Efficiency Nova Scotia. (n.d.). Guide to selecting a contractor. Retrieved from https://www.efficiencyns.ca/guide/guide-to-selecting-a-contractor/

[24] Public Utilities Act, RSNS 1989, c 380 at s. 79I.

[25] Ibid at s. 79C(2) and 79G(2).

[26] Ibid at s. 79J(1-2).

[27] Ibid at s. 79 J(3).

[28] Building Code Act, RSNS 1989, c. 46, ss. 1, 4.1 and 5.1.

[29] Nova Scotia Building Code Regulations, RSNS 1989, c. 46, N.S. Reg. 26/2017.

[30] Ibid, ss. 1.1.2.1(1).

[31] Ibid, ss. 1.1.1.2(2).

[32] Ibid, ss. 1.1.2.1(2).

[33] Ibid, ss. 1.1.1.2(2).
[34] Ibid, ss. 3.1.1.22, 3.1.1.32 and 3.1.1.23.
[35] Energy-efficient Appliances Act, SNS 1991, c. 2, ss. 2(c) and 3(1)(a).
[36] Energy-efficient Appliances Regulations, NS Reg. 400/2008.
[37] Ibid, ss. 2 and 8.
[38] Supra note 24.
[39] Municipal Government Act, SNS 1998, c 18, s 65 (aca) and 81A.

Chapter 9

[1] Nunavut Bureau of Statistics. (n.d.). Nunavut Quick Facts. Retrieved from http://www.stats.gov.nu.ca/en/home.aspx.
[2] Natural Resources Canada. (2016). Nunavut's Electric Reliability Framework. Retrieved from http://www.nrcan.gc.ca/energy/electricity-infrastructure/18840.
[3] Cherniak, D., Dufresne, V., Keyte, L., Mallett, A., Schott, S. (2015). Report on the State of Alternative Energy in the Arctic. School of Public Policy and Administration, Carleton University, Ottawa.
[4] Ibid.
[5] A Nothern Vision. (2016). Building a better north. Retrieved from http://www.anorthernvision.ca/.
[6] Governments of the Northwest Territories, Nunavut, and Yukon. (2011). Paths to a Renewable North: A Pan-territorial renewable energy inventory. Retrieved from http://www.anorthernvision.ca/documents/RenewableEnergyInventoryEN.pdf.
[7] Government of Nunavut. Department of Environment. (2011). Upagiaqtavut: Setting the Course: Climate Change Impacts and Adaptation in Nunavut. Retrieved from http://climatechangenunavut.ca/sites/default/files/3154-315_climate_english_reduced_size_1_0.pdf.
[8] Government of Nunavut. Department of Environment. (2016, November 4). GN announces creation of Climate Change Secretariat. Retrieved from http://gov.nu.ca/environment/news/gn-announces-creation-climate-change-secretariat.
[9] Government of Nunavut. Department of Environment. (n.d.). Climate Change Secretariat. Retrieved from http://www.gov.nu.ca/environment/information/climate-change-secretariat.
[10] Government of Nunavut. (2017). Climate Change Secretariat Strategic Plan 2017-2021. Retrieved from http://climatechangenunavut.ca/sites/default/files/2017-07-14-strategic_plan_ccs_fi-

nal_2.pdf.
[11] Ibid.
[12] Nunavut Energy. (n.d.). Climate Change Secretariat. Retrieved from http://www.nunavutenergy.ca/en/Energy_Secretariat.
[13] Nunavut Energy. (2007). IKUMMATIIT: The Government of Nunavut Energy Strategy. Retrieved from https://www.nunavutenergy.ca/sites/default/files/files/About%20Us%20Section/ikummatiit_energy_strategy_english.pdf.
[14] Nunavut Energy. (n.d.). Nunavut Housing Corporation. Retrieved from https://www.nunavutenergy.ca/en/node/209.
[15] Supra note 2.
[16] Supra note 13.
[17] Supra note 6.
[18] Beth Brown. (2017, July 6). Nunavut Senator says Canada's North needs a carbon tax exemption, for now. Nunatsiaq News. Retrieved from https://www.arcticnow.com/arctic-news/2017/07/06/nunavut-senator-says-canadas-north-needs-a-carbon-tax-exemption-for-now/.
[19] Legislative Assembly of Nunavut. (2008). HANSARD Official Report. 4th Sess., 2nd Ass. Retrieved from http://www.assembly.nu.ca/sites/default/files/Hansard_20080522.pdf.
[20] CBC News. (2007, November 9). Nunavut minister rejects committee's pan of light-bulb ban. CBC News. Retrieved from http://www.cbc.ca/news/canada/north/nunavut-minister-rejects-committee-s-pan-of-light-bulb-ban-1.677097.
[21] Nunavut Energy. (n.d.). Programs and Incentives. Retrieved from http://www.nunavutenergy.ca/Programs_and_Incentives.
[22] Natural Resources Canada. (2017). Home Renovation Program. Retrieved from http://oee.nrcan.gc.ca/corporate/statistics/neud/dpa/policy_e/details.cfm?searchType=default&s ectoranditems=all%7C0&max=10&pageId=1&categoryID=all&regionalDeliveryId=18&programTypes=4&keywords=&ID=4868.
[23] Nunavut Housing. (n.d.). Home Renovation Program. Retrieved from http://www.nunavuthousing.ca/hrp.

Chapter 10

[1] Mallinson, R. (2013, p. 145). Electricity Conservation Policy in Ontario: Assessing a System in Progress. Retrieved from http://sei.info.yorku.ca/files/2013/03/electricity-conservation-policy-on-

tario.pdf.

[2]   Norrie, S. J., Love, P. (2009). Creating a Culture of Conserva-
       tion in Ontario: Approaches, Challenges and Opportunities. Re-
       trieved from http://ieeexplore.ieee.org/document/5275904/?re-
       load=true.

[3]   Supra note 1, p. 148.

[4]   Supra note 2, p. 2.

[5]   Pratt, C., Boland, B. (2004, p. i.). Tough Choices: Addressing Ontar-
       io's Power Needs—Final Report to the Minister. Ontario, Canada:
       Electricity Conservation and Supply Task Force.

[6]   Ibid, p. ii.

[7]   Ibid, p. ii.

[8]   Ibid, p. 44.

[9]   Love, P. (2015, p. 33). Past, Present and Future of Energy Conser-
       vation in Ontario. Retrieved from http://www.energyregulation-
       quarterly.ca/articles/the-past-present-and-future-of-energy-con-
       servation-in-ontario#sthash.FyTfolD5.dpbs.

[10]   Electricity Act, S.O. (1998, c.15, s. 25.30(2)).

[11]   Ducan, D. (2006, July 13). Coordination and Funding of LDC ac-
       tivities to deliver Conservation and Demand-Side Management
       Programs [Letter to Dr. Jan Carr]. Toronto, Ontario. Retrieved from
       http://ieso.ca/-/media/files/ieso/document-library/ministeri-
       al-directives/2006/20060713-coordination-and-funding-of-ldc-ac-
       tivities-to-deliver-cdm-programs.pdf.

[12]   Supra note 1, p. 156.

[13]   Ontario Ministry of Energy. (2013). Conservation First: A Renewed
       Vision for Energy Conservation in Ontario. Retrieved from http://
       www.energy.gov.on.ca/en/files/2014/09/conservation-first-en.
       pdf.

[14]   Green Energy Act, 2009, S.O. (2009, c. 12). Retrieved from https://
       www.ontario.ca/laws/statute/09g12.

[15]   IESO. (2014). 2011-2014 Conservation Results Report. Retrieved
       from http://www.ieso.ca/-/media/files/ieso/document-li-
       brary/conservation-reports/annual/conservation-results-re-
       port-2011-2014.pdf.

[16]   Ibid, p. 3.

[17]   Ontario Ministry of Energy. (2013). Achieving Balance: Ontario's
       Long-Term Energy Plan. Retrieved from http://www.energy.gov.
       on.ca/en/files/2014/10/LTEP_2013_English_WEB.pdf

[18] Chiarelli, B. (2014, July 25). Industrial Accelerator Program [Letter to Colin Andersen]. Toronto, Ontario. Retrieved from http://ieso.ca/-/media/files/ieso/document-library/ministerial-directives/2014/20140725-industrial-accelerator-program.pdf.

[19] IESO. (2015). Charter for the Conservation First Implementation Committee (CIFC) and Associated Working Groups. Retrieved from http://ieso.ca/-/media/files/ieso/document-library/conservation/cfic/cfic-and-working-groups-charter.pdf -- page 1.

[20] Power Stream. (n.d.). POWER.HOUSE. Retrieved from https://www.powerstream.ca/innovation/power-house.html

[21] Environmental Commissioner of Ontario. (2009). Rethinking Energy Conservation in Ontario (vol. 1). Retrieved from https://eco.on.ca/reports/2009-energy-report-vol-1-rethinking-energy-conservation-in-ontario/

[22] IESO. (2016, December 16). IESO Announces Results of Demand Response Auction [News Release]. Retrieved from http://www.ieso.ca/en/corporate-ieso/media/news-releases/2016/12/ieso-announces-results-of-demand-response-auction.

[23] Ontario Energy Board. (2012). Renewed Regulatory Framework for Electricity. Retrieved from https://www.oeb.ca/industry/policy-initiatives-and-consultations/renewed-regulatory-framework-electricity.

[24] Ontario Energy Board. (2012). Information on regional infrastructure planning. Retrieved from https://www.oeb.ca/industry/tools-resources-and-links/information-regional-infrastructure-planning.

[25] Executive Council Directive 1411/2004 of June 2004. Retrieved from https://www.oeb.ca/documents/cases/RP-2004-0196/smartmeters_directiveJuly14_190704.pdf

[26] Government of Ontario. (2016, s.5). Climate Action Plan. Retrieved from https://www.ontario.ca/page/climate-change-action-plan#section-5.

[27] Supra note 13, p. 4.

[28] IESO. (2016). Ontario planning Outlook. Retrieved from http://www.ieso.ca/sitecore/content/ieso/home/sector-participants/planning-and-forecasting/ontario-planning-outlook.

[29] Ibid, Module 3, p. 9.

[30] Government of Ontario. (2017, December 15). The End of Coal. Retrieved from http://www.energy.gov.on.ca/en/archive/the-

end-of-coal/.

[31] Thibeault, G. (2017). Ontario's Energy Future. Lecture presented in Ontario's Château Laurier- 1 Rideau Street, Ottawa. http://www. economicclub.ca/events/display/ontarios-minister-of-energy.

[32] IESO. (2017, January 18). Ontario's Independent Electricity System Operator Releases 2016 Electricity Data. Retrieved from http://www.ieso.ca/en/corporate-ieso/media/news-releases/2017/01/ontarios-independent-electricity-system-operator-releases-2016-electricity-data.

[33] Bill 172: Climate Change Mitigation and Low-carbon Economy Act. (2016). Retrieved from http://www.ontla.on.ca/web/bills/bills_detail.do?locale=en&BillID=3740

[34] Government of Ontario. (2016). Climate Action Plan. Retrieved from https://www.ontario.ca/page/climate-change-action-plan.

[35] Government of Ontario. (2016). Green Investment Fund. Retrieved from https://www.ontario.ca/page/green-investment-fund.

[36] O. Reg. 46/17: ONTARIO CLIMATE CHANGE SOLUTIONS DEPLOYMENT CORPORATION. (2017). Filed February 17, 2017 under Development Corporations Act, R.S.O. 1990, c. D.10. Retrieved from https://www.ontario.ca/laws/regulation/r17046.

[37] Thibeault, G. (2017, February 15). [Letter to Bruce Campbell]. Toronto, Ontario. Retrieved from http://www.ieso.ca/-/media/files/ieso/document-library/ministerial-directives/2017/letter-minister-of-energy-occsdc-20170213.pdf?la=en.

[38] IESO. (2016, September 1). Ontario planning Outlook. Retrieved from http://www.ieso.ca/-/media/files/ieso/document-library/planning-forecasts/ontario-planning-outlook/ontario-planning-outlook-september2016.pdf page 2.

[39] IESO. (2016). Conservation Targets and Results. Retrieved from http://www.ieso.ca/en/sector-participants/conservation-delivery-and-tools/conservation-targets-and-results.

[40] Ontario Power Authority. (2015, p. vi.). Conservation First 2015-2020: EM&V Protocols and Requirements. Retrieved from http://ieso.ca/-/media/files/ieso/document-library/conservation/ldc-toolkit/emv-protocols-and-requirements-10312014.pdf.

[41] IESO. (2015, p. 7.). Conservation & Demand Management Energy Efficiency Cost Effectiveness Guide. Retrieved from http://ieso.ca/-/media/files/ieso/document-library/conservation/ldc-toolkit/cdm-ee-cost-effectiveness-test-guide-v2-20150326.pdf.

[42] Ibid, p. 9.
[43] Ibid, p. 8.
[44] Building Code Act, 1974, c. 74
[45] NRCan. (2013). Canada's national model codes development system. Retrieved from http://www.nrc-cnrc.gc.ca/eng/solutions/advisory/codes_centre/codes_brochure.html.
[46] Environmental Commissioner of Ontario. (2016, p. 63). Conservation: Let's Get Serious: Annual Energy Conservation Progress Report—2015/2016. Retrieved from http://docs.assets.eco.on.ca/reports/energy/2015-2016/ECO_Conservation_Lets_Get_Serious.pdf.
[47] Ibid, p. 81.
[48] Ibid, p. 79.
[49] O. Reg. 20/17: REPORTING OF ENERGY CONSUMPTION AND WATER USE. (2017). Filed February 6, 2017 under Green Energy Act, 2009, S.O. 2009, c. 12, Sched. A. Retrieved from https://www.ontario.ca/laws/regulation/r17020.
[50] Government of Ontario. (2017, October 17). Measure energy and water use for large buildings: How to report your building's water and energy consumption if it's 50 000 square feet of more. Retrieved from http://www.energy.gov.on.ca/en/ontarios-ewrb/.
[51] Government of Ontario. (2017, September 19). Manage energy costs for your business: For large to small businesses, manage your energy costs through one of these programs. Retrieved from http://www.energy.gov.on.ca/en/incentives-program-for-business/
[52] Government of Ontario. (2017, August 11). Municipal Energy Plan Program: Learn about funding to help your community improve its energy use. Retrieved from http://www.energy.gov.on.ca/en/incentives-programs-for-communities/).
[53] The Atmospheric Fund. (2017). About us. Retrieved from http://taf.ca/about-us/
[54] Efficiency Capital. (n.d.). Retrieved from http://efficiencycapital-corp.com/
[55] City of Toronto. (n.d.). Home Energy Loan Program. Retrieved from https://www.toronto.ca/services-payments/water-environment/environmental-grants-incentives-2/home-energy-loan-program-help/.
[56] Supra note 37.

[57] IESO. (n.d.). Education and Capacity Building Program: ECB Overview. Retrieved from http://www.ieso.ca/get-involved/funding-programs/education-and-capacity-building-program/overview.

[58] IESO. (n.d.). Aboriginal Community Energy Plans. Retrieved from http://www.ieso.ca/en/get-involved/funding-programs/aboriginal-community-energy-plans/overview.

[59] SMART GREEN. (2017). SMART Green Program Guidelines. Retrieved from https://cmeweb.crm.eperformanceinc.com/smart-green.

[60] City of Toronto. (n.d.). High-Rise Retrofit Improvement Program. Retrieved from http://www1.toronto.ca/wps/portal/contentonly?vgnextoid=e206c94d3dc4f410VgnVCM10000071d60f89RCRD.

[61] City of Guelph. (2016). Guelph Energy Efficiency Retrofit Strategy (GEERS). Retrieved from http://guelph.ca/plans-and-strategies/community-energy-initiative/geers/

[62] CMHC. (n.d.). On Reserve Housing—Retrofit Initiative. Retrieved from https://www.cmhc-schl.gc.ca/en/first-nation/financial-assistance/renovation-programs/upload/on-reserve-housing-retrofit-initiative.pdf

[63] Hydro One. First Nations Conservation Program. Retrieved from https://www.hydroone.com/saving-money-and-energy/residential/first-nations-conservation-program.

[64] Ontario Aboriginal Housing Services. (2017). Programs. Retrieved from http://www.ontarioaboriginalhousing.ca/programs/.

[65] Ontario Energy Board. (2014, p.1.). Demand-Side Management Framework for Natural Gas Distributors (2015-2020) (Report of the Board EB-2014-0134). Retrieved from https://www.oeb.ca/oeb/_Documents/EB-2014-0134/Report_Demand_Side_Management_Framework_20141222.pdf.

[66] Ontario Energy Board. (1993). A Report On The Demand-Side Management Aspects Of Gas Integrated Resource Planning For: The Consumers' Gas Company LTD. Centra Gas Ontario INC. And Union Gas Limited (E.B.O. 169-III). Retrieved from https://www.oeb.ca/documents/cases/EB-2006-0021/EBO_169_III.pdf.

[67] Ontario Energy Board. (2006, August 25). Decision with Reasons (EB-2006-0021). Retrieved from https://www.oeb.ca/documents/cases/EB-2006-0021/dec_dsm_250806.pdf.

[68] Ontario Energy Board. (2011, June 30). Demand-Side Management

Guidelines for Natural Gas Utilities (EB-2008-0346). Retrieved from https://www.oeb.ca/oeb/_Documents/Regulatory/DSM_Guidelines_for_Natural_Gas_Utilities.pdf.

[69] Supra note 67, p. 2.

[70] Ontario Executive Council. (2014, March 26). Minister's Directive to the Ontario Energy Board. Retrieved from https://www.oeb.ca/oeb/_Documents/Documents/Directive_to_the_OEB_20140326_CDM.pdf

[71] Supra note 67, p. 2.

[72] Supra note 67, p. 5.

[73] Supra note 67, p. 14.

[74] Ministry of the Environment and Climate Change. (2016, May 19). Ontario Posts Final Cap and Trade Regulation: Province Reducing Greenhouse Gas Pollution, Creating Jobs [News Release]. Retrieved from https://news.ontario.ca/ene/en/2016/05/ontario-posts-final-cap-and-trade-regulation.html.

[75] Office of the Premier. (2016, June 8). Ontario Releases New Climate Change Action. Retrieved from https://news.ontario.ca/opo/en/2016/06/ontario-releases-new-climate-change-action-plan.html.

[76] Supra note 17.

[77] Supra note 67 p. 17.

[78] Supra note 67 p. 20.

[79] Supra note 67 p. 30.

[80] Supra note 67 p. 33.

[81] O. Reg. 397/11: ENERGY CONSERVATION AND DEMAND MANAGEMENT PLANS (2009). Filed under Green Energy Act, 2009, S.O. 2009, c. 12, Sched. A. Retrieved from https://www.ontario.ca/laws/regulation/110397.

[82] Ontario Ministry of Energy. (2017). Conservation for Public Agencies. Retrieved from http://www.energy.gov.on.ca/en/green-energy-act/conservation-for-public-agencies/.

[83] Supra note 54.

Chapter 11

[1] Statistics Canada. (2017). Supply and demand of primary and secondary energy in terajoules. (CANSIM 128-0016). Retrieved from http://www5.statcan.gc.ca/cansim/a26?lang=eng&id=1280016.

[2] Ibid.

[3] PEI Department of Environment and Energy. (2004). Energy Framework and Renewable Energy Strategy. Retrieved from http://www.gov.pe.ca/photos/original/ee_frame_rep_e.pdf.

[4] PEI Department of Environment, Energy and Forestry. (2008). Prince Edward Island Energy Strategy, Securing our Future: Energy Efficiency and Conservation. Retrieved from http://www.gov.pe.ca/photos/original/env_snergystr.pdf.

[5] PEI Energy Commission. (2012). Final Report: Charting Our Electricity Future. Retrieved from http://www.gov.pe.ca/photos/original/NRGCommish_13.pdf.

[6] Government of Prince Edward Island. (2017). PEI PROVINCIAL ENERGY STRATEGY 2016/17. Retrieved from https://www.princeedwardisland.ca/sites/default/files/publications/pei_energystrategymarch_2017_web.pdf.

[7] Supra note 5.

[8] Supra note 8.

[9] Government of Prince Edward Island. (2016). Efficiency PEI. Retrieved from https://www.princeedwardisland.ca/en/information/transportation-infrastructure-and-energy/efficiencypei.

[10] Supra note 6.

[11] Supra note 5.

[12] Government of Canada. (n.d.). Efficiency PEI. Retrieved from http://oee.nrcan.gc.ca/corporate/statistics/neud/dpa/policy_e/results.cfm?programtypes=4&regionaldeliveryid=3&attr=0

Chapter 12

[1] Ministry of Energy and Natural Resources. (n.d.). Du Bureau des économies d'énergie à l'Agence de l'efficacité énergétique. Retrieved from http://archive.wikiwix.com/cache/?url=http%3A%2F%2Fwww.aee.gouv.qc.ca%2Flagence%2Fhistorique%2F.

[2] Langlois, P., Hansen, S. (2012). World ESCO Outlook. Lilburn, GA : Fairmont Press.

[3] Hydro-Québec Distribution. (2002). Bref historique des interventions d'Hydro-Québec en efficacité énergétique. Retrieved from http://www.regie-energie.qc.ca/audiences/3473-01/HQ/HQD2-1_5nov02.pdf.

[4] Ministère des Ressources naturelles. (1996). L'énergie au service du Québec : Une perspective de développement durable. Retrieved from http://mern.gouv.qc.ca/energie/politique/pdf/

Strategie%20Energie%201996.pdf.

[5]   Agence de l'efficacité énergétique. (2006). AEE 1997-2006 Activity Reports.

[6]   Loi sur l'Agence de l'efficacité énergétique, c. 46, a. 14, 2006.

[7]   Agence de l'Efficacité Énergétique. (2008). AEE 2007-2008 Activity Report.

[8]   Agence de l'Efficacité Énergétique. (2009). AEE 2008-2009 Activity Report.

[9]   Ministère du Développement durable, de l'Environnement et de la Lutte contre les changements climatiques. (n.d.). Fonds vert. Retrieved from http://www.mddelcc.gouv.qc.ca/ministere/fonds-vert/.

[10]   Transition énergétique Québec. (n.d.). Programs and financial assistance. Retrieved from http://www.transitionenergetique.gouv.qc.ca/en/programs-and-financial-assistance/page-pro-grammes/1/?tx_nurprogsubv_pi1%5Benergy%5D=24&tx_nur-progsubv_pi1%5Border%5D=energy#.WleDCqjiaM8.

[11]   Régie de l'énergie Québec. (n.d.). La Régie de l'Énergie. Retrieved from http://www.regie-energie.qc.ca/

[12]   Ministère de l'Énergie et des Ressources naturelles. (2015). Politique énergétique 2016-2025—Efficacité et innovation énergétiques (Fascicule 3). Retrieved from https://mern.gouv.qc.ca/energie/politique/documents/fascicule-3-BEIE.pdf.

[13]   Gouvernement du Québec. (2016). Décret 746-2016. Gazette officielle du Québec. Retrieved from http://www2.publications-duquebec.gouv.qc.ca/dynamicSearch/telecharge.php?type=1&file=65439.pdf.

[14]   Hydro-Québec. (n.d.). Sustainable Development. Retrieved from http://www.hydroquebec.com/sustainable-development/energy-environment/global-plan.html.

[15]   Hydro-Québec Distribution. (2013). Plan d'approvisionnement 2014-2023 Réseau intégré. Retrieved from http://publicsde.regie-energie.qc.ca/projets/232/DocPrj/R-3864-2013-B-0005-Demande-Piece-2013_11_01.pdf.

[16]   Hydro-Québec Distribution. (2015). Interventions en efficacité énergétique. Retrieved from http://publicsde.regie-energie.qc.ca/projets/317/DocPrj/R-3933-2015-B-0042-Demande-Piece-2015_07_30.pdf.

[17]   Gaz Métro. (2016). Gaz Métro : prête pour les défis à l'horizon

2030—Fiche d'information. Retrieved from https://www.ener-
gir.com/~/media/Files/Corporatif/Fiches/Fiche_Transport_
avril2016.pdf?la=fr.

[18] Gaz Métro. (2017). Plan global en efficacité énergétique—Horizon 2017-2019, Cause tarifaire, Retrieved from http://publicsde.
regie-energie.qc.ca/projets/372/DocPrj/R-3970-2016-B-0209-
DemAmend-PieceRev-2016_09_01.pdf.

[19] Supra note 12.

[20] Quebec Government. (2016). The 2030 Energy Policy—Energy in
Quebec: A Source of Growth. Retrieved from https://politiqueener-
getique.gouv.qc.ca/wp-content/uploads/Energy-Policy-2030.pdf.

[21] Ministère de l'Énergie et des Ressources naturelles. (2017). 2030
Energy Policy Action Plan. Retrieved from https://politiqueen-
ergetique.gouv.qc.ca/wp-content/uploads/Tableau-PA-PE2030_
ANG.pdf.

[22] Government of Quebec. (2012). Quebec in Action Greener by 2020
in 2013-2020 Climate Change Action Plan. Retrieved from http://
www.mddelcc.gouv.qc.ca/changements/plan_action/pacc2020-
en.pdf.

[23] Canadian Intergovernmental Conference Secretariat. (2015).
Resolution 39-1 Concerning Climate Change (39th Annual Conference of NEG/ECP). Retrieved from http://www.scics.ca/
en/product-produit/resolution-39-1-resolution-concerning-cli-
mate-change/

[24] Ministère du Développement durable, de l'Environnement et de
la Lutte contre les changements climatiques. (2016). Inventaire
québécois des émissions de gaz à effet de serre en 2014 et leur évo-
lution depuis 1990. Retrieved from http://www.mddelcc.gouv.
qc.ca/changements/ges/2014/Inventaire1990-2014.pdf.

[25] Gouvernement du Québec. (2015). Banque de données des statis-
tiques officielles sur le Québec. Retrieved from http://www.
bdso.gouv.qc.ca/pls/ken/ken214_tabl_detl.page_detl?p_iden_
tran=REPERMAU4RY09-140823886859bRjq&p_lang=1&p_id_
raprt=1412.

[26] Ministère du Développement durable, de l'Environnement et de
la Lutte contre les changements climatiques. (2013). A Brief Look
at the Quebec Cap-and-Trade System for Emission Allowances.
Retrieved from http://www.mddelcc.gouv.qc.ca/changements/
carbone/ventes-encheres/spede-enbref-en.pdf.

[27] Agence de l'Efficacité Énergétique. (2009). AEE 1998-2009 Activity Reports.
[28] Supra note 12.
[29] Quebec Government, "The 2030 Energy Policy—Energy in Quebec: A Source of Growth," 2016.
[30] Supra note 12.
[31] Molina, M. (2014). The Best Value for America's Energy Dollar: A National Review of the Cost of Utility Energy Efficiency Programs. (Research Report U1402). Retrieved from http://aceee.org/research-report/u1402.
[32] National Energy Board. (2017). Canada's Adoption of Renewable Power Sources—Energy Market Analysis. Retrieved from https://www.neb-one.gc.ca/nrg/sttstc/lctrct/rprt/2017cnddptnrnwblpwr/cststrdffs-eng.html?=undefined&wbdisable=true.
[33] Supra note 12.
[34] RBC Royal Bank. (n.d.). RBC Energy-Saver™ Loan. Retrieved from http://www.rbcroyalbank.com/personal-loans/energy-saver-loan.html.
[35] Desjardins. (n.d.). Energy-Efficiency Loan. Retrieved from https://www.desjardins.com/ca/business/financing-credit/long-term-financing/energy-efficiency-loan/index.jsp.

Chapter 13
[1] Statistics Canada. (2017). Supply and demand of primary and secondary energy in terajoules. (CANSIM 128-0016). Retrieved from http://www5.statcan.gc.ca/cansim/a26?lang=eng&id=1280016.
[2] VanEE. (n.d.). About us. Retrieved from http://www.vanee.ca/en/about-us.html.
[3] University of Regina Press. (n.d.). Energy efficient houses.
[4] Saskatchewan Energy Conservation and Development Authority. (1994). Technical Assessment of the Nuclear Option for Saskatchewan. Retrieved from http://www.saskpower.com/wp-content/uploads/1994_technical_assessment_nuclear_for_SK.pdf.
[5] Dumont, R., Morin. T. (n.d.). The factor 9 home: a new approach. Retrieved from http://riverdalenetzero.ca/REPORTS/The_Factor_9_Home_--_A_New_Prairie_Approach.pdf.
[6] Dumont, R. (2008). The Factor 9 Home: Reducing Energy Consumption by 90% [PowerPoint Presentation]. Retrieved from http://www.emtfsask.ca/presentations/saskatoon/08-10-01-RobDu-

mont.pdf.

[7]   Peter, N. (2013, October 1). Factor 9 home. SRS Research Council. Retrieved from http://www.src.sk.ca/blog/factor-9-home.

[8]   Huck, N. (2015, August 5). 'Passive home' movement a success in Germany but not in Saskatchewan where it started. CBC News. Retrieved from http://www.cbc.ca/news/canada/saskatche-wan/passive-home-movement-a-success-in-germany-but-not-in-saskatchewan-where-it-started-1.3179851.

[9]   Dodge, D., Thompson, D. (2016, May 6). The first certified passive house in Saskatchewan [Blog Post]. Retrieved from http://www.pembina.org/blog/first-certified-passive-house-saskatchewan.

[10] SaskPower (2014). SaskPower Industrial Energy Optimization Program: Program Guide (v. 2.2). Retrieved from http://www.saskpower.com/wp-content/uploads/ieop_program_guide.pdf.

[11] SaskEnergy. (n.d.). Commercial Boiler Program. Retrieved from http://www.saskenergy.com/business/commercialboiler.asp.

[12] SaskPower (2017). Reseller Rates. Retrieved from http://www.saskpower.com/wp-content/uploads/Service_Rates_Re-seller_2017.pdf.

[13] City of Saskatoon Transportation & Utilities Department. (2015). 2015 Annual Report. Retrieved from https://www.saskatoon.ca/sites/default/files/documents/transportation-utilities/saska-toon-light-power/2015_annual_report.pdf.

[14] SES Solar Co-operative Ltd. (2015). Where it all began. Retrieved from https://sessolarcoop.wildapricot.org/about-us.

[15] Tank, P. (2017). Car-share pilot project powered by solar. Saskatoon Star Phoenix. http://thestarphoenix.com/news/local-news/car-share-pilot-project-powered-by-solar.

[16] Sun Country Highway. (2017). Retrieved from https://suncoun-tryhighway.com.

[17] Morrell, K. (n.d.). Wind of Change at Saskatoon's WDM. Western Development Museum Saskatoon. Retrieved from http://www.wdm.ca/stoon/wind_extra.html.

[18] Western Development Museum Saskatoon. (2017). Fuelled by Innovation. Retrieved from http://wdm.ca/stoon/innovation.html.

[19] Wolsfeld, M. (2015, March 24). Energy Conservation Month Highlight: Lighting Retrofits. University of Saskatchewan Office of Sustainability. Retrieved from http://sustainability.usask.ca/news/2015/energy-conservation-month-highlight-lighting-retro-

fits.php.

[20] University of Saskatchewan Office of Sustainability (n.d.). Energy Consumption. Retrieved from http://sustainability.usask.ca/footprint/energy.php#Initiatives.

[21] City of Saskatoon. (n.d.). Clean Power Generation Initiatives. Retrieved from https://www.saskatoon.ca/services-residents/power-water/saskatoon-light-power/clean-power-generation-initiatives.

[22] Government of Saskatchewan. (2003). Saskatchewan mandates energy efficiency in public buildings. Retrieved from http://www.saskatchewan.ca/government/news-and-media/2003/december/15/saskatchewan-mandates-energy-efficiency-in-public-buildings.

[23] Saskatchewan Urban Municipalities Association. (n.d.). Kelln solar. Retrieved from http://suma.org/sumadvantage/detail/solar-pool-heating-kelln-solar.

[24] Kelln Solar. (n.d.). Solar Pool Heating. Retrieved from http://kellnsolar.com/products/n?n=51.

[25] Government of Saskatchewan. (n.d.). National Building Code and National Fire Code Information. Retrieved from https://www.saskatchewan.ca/business/housing-development-construction-and-property-management/building-standards-and-licensing/national-building-and-fire-code-information.

[26] F-13.4 Reg 32: The Energy-Efficient Household Appliances (provincial Sales Tax) Remission and Exemption Regulations. (2005). Publications Saskatchewan. Retrieved from http://www.publications.gov.sk.ca/details.cfm?p=10285.

[27] Government of Saskatchewan. (2003). The Provinvial Salex Tax Act. Information Bulletin. Retrieved from http://finance.gov.sk.ca/revenue/pst/bulletins/PST069EnergyEfficientAppliances.pdf.

[28] Honeywell. (2010, March 25). SaskPower/Honeywell Help Saskatchewan Businesses Save Energy And Money. Retrieved from http://www51.honeywell.com/honeywell/news-events/press-releases-details/03.25.10SaskPowerandHoneywell.html.

[29] Government of Saskatchewan. (2013). Go Green Fund. Retrieved from http://www.environment.gov.sk.ca/go-green/fund.

[30] SaskEnergy. (2007, June 25). Energy Efficient Rebate Program Announced for New Homes. Retrieved from http://www.saskener-

gy.com/About_SaskEnergy/News/news_releases/2007/070626.
asp.

[31] SaskPower. (n.d.). Business Programs and Offers. Retrieved from
      http://www.saskpower.com/efficiency-programs-and-tips/busi-
      ness-programs-and-offers/.

[32] SaskEnergy. (n.d.). Energy Efficiency Programs. Retrieved from
      http://www.saskenergy.com/saving_energy/specialoffers.asp.

[33] SaskEnergy. (n.d.). The ENERGY STAR® Loan Program. Retrieved
      from http://www.saskenergy.com/residential/appliancefinanc-
      ing.asp.

[34] Affinity Credit Union. (2017). Specialty Financing. Retrieved from
      https://www.affinitycu.ca/Community/Lending/Pages/Retrof-
      itLoans.aspx.

[35] Ibid.

[36] The Cities Act of the Statutes of Saskatchewan (2002, C-11.1). Re-
      trieved from http://www.publications.gov.sk.ca/freelaw/docu-
      ments/english/Statutes/Statutes/c11-1.pdf.

[37] The Local Improvements Act of the Statutes of Saskatchewan
      (1993, L-33.1). Retrieved from http://www.publications.gov.sk.
      ca/freelaw/documents/English/Statutes/Statutes/L33-1.pdf.

Chapter 14

[1]   Yukon Government. (2017). Programs. Retrieved from http://
      www.energy.gov.yk.ca/programs.html accessed May 2017.

[2]   In Charge Yukon. (2017). Retrieved from http://www.inchargeyu-
      kon.ca/ accessed May 2017.

[3]   Yukon Housing Corporation. (n.d.). Home Repair Program.
      http://www.housing.yk.ca/pdf/loanYH16040factsheet_Loan-
      Portfolio_HRL_web.pdf accessed May 2017.

Chapter 15

[1]   Natural Resources Canada. (2013). Energy Performance Contract-
      ing: Guide for Federal Buildings. Retrieved from https://www.
      nrcan.gc.ca/sites/www.nrcan.gc.ca/files/oee/files/pdf/com-
      munities-government/buildings/federal/pdf/12-0419%20-%20
      EPC_e.pdf.

[2]   Langlois, P., Hansen, S. (2012). World ESCO Outlook. Lilburn, GA:
      Fairmont Press Inc.

[3]   Natural Resources Canada. (2011). Communities and Government:

Federal Buildings Initiative. Retrieved from http://oee.nrcan.gc.ca/communities-government/buildings/federal/federal-buildings-initiative.cfm.

[4] Energy Services Association of Canada. (2017). Guaranteed Energy Savings. Retrieved from http://energyservicesassociation.ca/documents/ESAC-GES-Issue1-2017.pdf.

[5] Energy Services Association of Canada. (n.d.). Retrieved from www.energyservicesassociation.ca.

[6] Natural Resources Canada. (n.d.). Case Studies. Retrieved from http://www.nrcan.gc.ca/commercial/cbr/pubs/4201.

[7] Energy Services Association of Canada. (n.d.). Case Studies & Industry Profiles. Retrieved from http://energyservicesassociation.ca/case-studies/index.html.

[8] Energy Services Association of Canada. (2016). White Paper on the Use of energy Service Performance Contracts (ESPCs) to Achieve Provincial Carbon Reduction Targets. Retrieved from http://energyservicesassociation.ca/documents/White-Paper-2016Apr28-ON.pdf.

[9] Ontario Government. (2016). Climate Change Action Plan. Retrieved from https://www.ontario.ca/page/climate-change-action-plan.

[10] Law, A. (2010). Energy Efficiency within Alberta Infrastructure Buildings [PowerPoint Presentation]. Alberta Energy Efficiency Association Conference. Retrieved from http://www.aeea.ca/files/nov4event/Alberta%20Infrastructure.pdf.

[11] Alberta Energy Efficiency Advisory Panel. (2016). Getting it Right: A More Energy Efficient Alberta. Retrieved from https://www.alberta.ca/documents/climate/EEAP-Report-Getting-It-Right-Complete.pdf.

[12] Gauthier, Genevieve. (2016). Energy Performance Contracting: Key Considerations to Maximize Benefits [PowerPoint Presentation]. Retrieved from http://www.rpic-ibic.ca/documents/2016_RP_NW/Presentations/201611_RPIC_Econoler_v1_1.pdf.

Index

Symbols

21st Conference of the Parties (COP21) 198

2007 Energy Plan 36, 37

2010 Clean Energy Act (CEA) 44

2011 NWT Greenhouse Gas Strategy 95

2013 LTEP 150

2013 NWT Energy Action Plan 96

2014/15-2016/17 Electricity Efficiency Plan 72

2014-2023 Supply Plan (Plan d'approvisionnement 2014-2023) 194

2015-2020 Conservation First Framework (CFF) 136, 139

2015-2020 DSM Framework 167, 171

2016 Climate Change Action Plan: Transitioning to a Low-Carbon Economy 67

2016 Climate Leadership Plan 37

2016 Energy Policy 195

2016 FortisBC (electric) LTRP 39

2030 Energy Policy 197, 199

A

Aboriginal Community Energy Plan Program 164

Act Respecting the Régie de l'Énergie 201

Act Respecting Transition Énergétique Québec 201

administrative models 105

Advanced Building Recommissioning (RCx) 248

AEA 96, 103

programs 100, 102

Agreement on the Transfer of Federal GTF 93

Alberta 2015 Climate Leadership Plan 28

Alberta Climate Change Office (ACCO) 26

Alberta Energy Efficiency Advisory Panel 237

Alberta Energy Efficiency Alliance 27

Alberta Infrastructure 237

Alternative Energy Technologies Program 100

Alternative Energy Technologies Program (AETP) 103

annual duty 192

Arctic Energy Alliance 93, 98

ASHRAE 90 Standard 14

ATCO Gas 25

Atlantic Energy Framework for Collaboration 177

audits 152

B

barriers 16, 46, 270

BC Building Code 45

BC EE programs 35

BC Energy Efficiency Act 32

BC Energy Plan—A Vision for Clean Energy Leadership 35

BC Hydro 32

BC Ministry of Energy and Mines (MEM) 34

B.C.'s carbon neutral government

program 238
BEIE 190
Biomass Energy Fund (Fonds Bio-
    masse Énergie) 208
BOMA BEST 28
British Columbia Utilities Com-
    mission (BCUC) 38
broader public sector (BPS) 159
Build Better Buildings (BBB) Pol-
    icy 82
Building Act 44
Building Code (2012) 159
Building Code Regulations 119
building codes 34, 150
Building Energy Retrofit Program
    235
Building Renewal Program 235
Bureau de l'efficacité énergétique
    (BEE, Energy Efficiency Bu-
    reau) 183
Bureau of Energy Efficiency and
    Innovation (Bureau de l'ef-
    ficacité et de l'innovation
    énergétiques—BEIE) 189
Business, Non-Profit and Institu-
    tional (BNI) Energy Savings
    Program 26

C
Canada Border Services Agency
    (CBSA) 9
Canada Energy Strategy 10
Canada Infrastructure Bank 21
Canada Mortgage and Housing
    Corporation (CMHC) 9
Canada Revenue Agency (CRA) 9
Canadian Commission on Build-
    ing and Fire Codes 9
Canadian Energy Efficiency Alli-

ance (CEEA) 253
Canadian Home Builders' Associ-
    ation (CHBA) 9
Canadian Industry Program for
    Energy Conservation
    (CIPEC) 4
Canadian Institute for Energy
    Training (CIET) 247
CanmetENERGY 8
CAN-QUEST 248
Capacity-Based Demand Re-
    sponse (CBDR) program 145
cap-and-trade program 115, 153,
    170, 236
cap-and-trade system 181, 202,
    203, 217
Capital Asset Retrofit Fund
    (CARF) 99
carbon capture and sequestration
    (CCS) 216
carbon emissions 269
carbon levy 25, 26
carbon price mechanism 71
carbon pricing 99, 115, 129
carbon tax 58, 60, 181, 217
Carbon Tax Act 46
carbon tax mechanism 99
Carbon Tax Regulation 46
CCA 19
CCNL 80
CDM pilot projects 163
CDM plans 139, 160
CDM programs 134, 136, 148
    in Ontario (2015-2020) 137
CEM 247
CEMET (Canadian Energy Man-
    agement and Environmental
    Training) 245
Centra Gas Manitoba 56

certification program 69, 187
City of Medicine Hat's utility system 27
City of Swift Current 213
City of Vancouver 41
Clean Energy Act 37
Clean Foundation 110
climate change 11, 21, 79, 83, 100, 115, 125, 129, 151, 164, 177, 199, 254
Climate Change Action Plan 152, 236
Climate Change and Emissions Management Fund (CCEMF) 25
Climate Change Central 24
Climate Change Memorandum of Understanding (MOU) 175
Climate Change Mitigation and Low-Carbon Economy Act 153
Climate Change Secretariat Strategic Plan 2017-2021 126
Climate Leadership Plan 25
CMHC Green Home Program 20
combined heat and power (CHP) 215
Commercial Buildings Retrofit Program 72
Commercial Energy Audit Program 182
Commercial Energy Conservation and Efficiency Program (CE-CEP) 103
Commercial Lighting Retrofit Program 165
commissioning 234
Community Building Energy Retrofit Program) 103

Community Government Building Energy Retrofit Program 100
comparative labels 5
Conservation Corps of Newfoundland and Labrador 75
Conservation First Framework 249
Conservation First Implementation Committee (CFIC) 140
conservation potential review (CPR) 225
cost effectiveness 157
    metrics 158
custom retrofit 120

D
Data Strategy Advisory Council 148
Deal Days point-of-sale discount program 154
demand response programs 144
demand side 54
    approach 53
    management 18, 59
Demand/Supply Plan 132
Department of Community and Government Services (CGS) 126
Department of Municipal Affairs and Environment 79
Dollars to $ense 245, 249
Dollars to $ense Energy Management Workshops 12
DSM 31, 36
    plan 72, 181
    programs 25, 39, 107, 168, 186, 223, 230, 243
    regulation 43
    strategy 55

E
ecoEnergy Retrofit Program 81
Econoler 230
Econoler ESCO market 230
Econoler Inc. 185
Education and Capacity Building
    Program 164
energy efficiency (EE) 11, 57, 114,
    185, 215, 222
    assessment 17
    barriers 126
    gains 271
    in building codes 14, 102
    potential 77
    programs 109
    regulations 13
    requirements 148
    retrofit loans 219
    standards 60
    strategy 82, 186
    training 246
        programs 248
Efficiency Capital (EC) 164
Efficiency Manitoba Act—2017 59
Efficiency Matters 256
Efficiency Nova Scotia (ENS) 107
EfficiencyOne 114, 117, 234
    Services 121
Efficiency PEI 178
Efficiency Services Division of NB
    Power 66
electrical energy and demand sav-
    ings 112, 116
Electricity Conservation & Supply
    Task Force (ECSTF) 133
Electricity Efficiency and Conser-
    vation Restructuring (2014)
    107
Electricity Efficiency Plan 66

electricity plans 146
electricity savings 187
endorsement labels 5
Énergir 192, 195
Énergir EE programs 196
EnerGuide 4, 17
Energy and Climate Partnership
    of the Americas (ECPA) 7
energy and demand savings pro-
    grams 260
Energy and Minerals Division 178
energy charrette 96
Energy Competition Act 132
Energy Conservation Leadership
    Act 135
energy conservation programs 77
energy distributors 127
energy DSM 97
Energy Efficiency Act (EEA) 4,
    13, 42
Energy Efficiency Agency
    (l'Agence de l'efficacité
    énergétique, AEE) 187
Energy Efficiency Alberta 25, 26
Energy Efficiency Incentive Pro-
    gram (EEIP) 103
Energy Efficiency Regulations 5
Energy Efficiency Standards Reg-
    ulation (EESR) 44
Energy-efficient Appliances Act
    119
Energy-efficient Appliances Regu-
    lations 119
energy management 246
    certifications 247
Energy Manager Initiative 156
energy performance contracting
    (EPC) 20, 185
Energy Performance Program

(EPP) 156
energy policy 65, 187
Energy Priorities Framework
    (2008) 101
energy savings 110
Energy Savings Performance
    Agreement (ESPA) 271
energy service companies (ES-
    COs) 20, 229
Energy Services Association of
    Canada 232
Energy Serving Quebec: A Sus-
    tainable Development
    Perspective (L'énergie au
    service du Québec: Une per-
    spective de développement
    durable) 187
ENERGY STAR® 17, 18, 194, 224
    LED bulbs 225
    Loan Program 219
    Portfolio Manager 161, 248
Energy Step Code 47
Energy utilities 35
ENERGY WISE program 194
EnerQuality 255
ENS 108
    Trade Partner Network 116
Environment and Climate Change
    Canada (ECCC) 10
EPC projects 218
    in Canada 231
EPCs 28
ESCO 185
evaluation, measurement and ver-
    ification 157, 171

F
Factor 9 Home 212
FBI 20, 238

program 232
Federal Buildings Initiative (FBI)
    5
Federal Sustainable Development
    Act 11
Federal Sustainable Development
    Strategy (FSDS) 233
Federation of Canadian Munici-
    palities (FCM) 10
financial incentive 20, 72
Five-Year Conservation Plan:
    2016-2020 76, 80
Fondaction 208

G
Gas DSM 172
    Funding 170
Gazifère 192
General Regulation—Energy Effi-
    ciency Act 71
German PassivHaus program 212
GHG emission 11, 35, 69, 99, 161,
    162, 163, 165, 170, 198, 233,
    238, 272
    reductions 267
    targets 151, 162
Global Energy Efficiency Plan
    (Plan global en efficacité
    énergétique) 195
GN energy strategy 127
Go Green Fund 218
Good Building Practice for North-
    ern Facilities guide 100
Good Energy Program 223
Government of Newfoundland
    and Labrador 76
Green Action Centre 57
Green Buildings BC Retrofit Pro-
    gram 238

Green Condo Loan 271
Green Energy Act 135, 159, 160
green fund 153, 190
Green Home Program 6
Greenhouse Gas (GHG) 46
Greenhouse Gas Strategy 2011-2015 100
Greening Government Action Plan 83
Green Investment Fund (GIF) 164
Green Municipal Fund (GMF) 21
Guelph Energy Efficiency Retrofit Strategy (GEERS) 165

H
harmonizing standards 16
HAT Smart program 28
High-rise Retrofit Improvement Support Program (Hi-RIS) 165
Home Energy Efficiency Loan Program (HEELP) 90
Home Energy Savings Program (HESP) 89
Home Insulation Energy Savings Program 72
Home Repair Loan program 226
Hydro-Québec 187, 192, 193

I
IESO 139
    Conservation Fund 141
    Market Renewal program 145
    Ontario Planning Outlook 150
impact evaluation 242
incentive programs 47, 163
incentives 75
inCharge 225

incremental energy 116
    savings 140
independent facilitators 238
Industrial Accelerator Program (IAP) 139
Industrial Conservation Initiative 143
industrial programs 66
Innovative Clean Energy (ICE) Fund 32
Integrated Power System Plan 134
intergovernmental collaborations 9
International Partnership for Energy Efficiency Cooperation (IPEEC) 6
International Performance Measurement and Verification Protocol (IPMVP) 243
Investment Canada Plan 21
IO portfolio 163
IRP 54
Island Interconnected System 76
Island Regulatory and Appeals Commission (IRAC) 179
ISO 50001 Energy Management Standard 268

L
labeling process 17
labeling programs 16
labeling standards 201
La Stratégie Québécoise d'efficacité énergétique: une contribution au développement durable 186
LDC-delivered programs 154
LDC-led program design 143
Leadership in Energy and Envi-

ronmental Design (LEED) Program 82, 222
LEED Silver 28
Certification 215
Local Government (Green Communities) Statutes Act 46
local planning 147
low-carbon economy 170, 197, 256
low carbon footprint 199
low-carbon solutions 270
lower-carbon global economy 79
Low-income Energy Savings Program 72

M

Manitoba branch of the Consumers' Association of Canada (CAC Manitoba) 57
Manitoba Climate Change and Green Economy Action Plan 58
Manitoba EE policy 57
Manitoba Energy Code for Buildings (MECB) 59
Manitoba Geothermal Energy Incentive Program 61
Manitoba Hydro 53
key focus areas 56
Manitoba Public Utilities Board (PUB) 56
Maritime Electric 179
market barriers 33, 65, 153
market evaluation 242
market transformation 33, 141, 186, 273
framework 83
measurement and verification (M&V) 234, 243, 268

MEPS 44, 159
Meter Data Management/Repository (MDM/R) 147
minimum energy performance standards 15, 60
Ministerial Energy Coordinating Committee (MECC) 97
Ministry of Energy 161
Modified Total Resource Cost 38
MOI 162
Moving Forward: Energy Efficiency Action Plan 82
Municipal Energy Plan (MEP) program 172

N

National Building Code 181
of Canada 84
National Energy Code of Canada for Buildings (NECB) 14
National Energy Use Database (NEUD) 5
natural gas DSM stakeholders 169
Natural Resource Council of Canada (NRC) 34
NBCC 119
requirements 84
NB Power Corporation (NB Power) 64, 67
NECB 119
Net Present Value (NPV) 234
net-zero carbon emission homes 149
net-zero construction 246
net-zero energy 33, 212
NetZero Home 212
NEUD 18
New Brunswick Department of

Energy and Mines 67
New Brunswick Energy Blueprint
    66, 69
new construction 44, 45, 66, 120,
    267
Newfoundland and Labrador
    Board of Commissioners of
    Public Utilities (PUB) 84
Newfoundland and Labrador En-
    vironmental Industry Asso-
    ciation (NEIA) 80
Newfoundland & Labrador Hy-
    dro (NFL Hydro) 76
Newfoundland Power 76, 78
NFL Hydro 78
Northern Premiers' Forum 125
Northwest Territories (NWT) 97
Northwest Territories Power Cor-
    poration (NTPC) 98
Nova Scotia's Building Code Act
    118
NRCan 6, 8, 17, 34, 182, 246
NS Power 113
Nunavut Energy Centre (NEC)
    127
Nunavut Energy Retrofit Program
    (NERP) 125
Nunavut Housing Corporation
    (NHC) 127
NWT Energy Action Plan 98
NWT Power System Plan 98

O
OCC 79, 83
OEB 167
Office of Energy Efficiency 8
Office of Energy Research and De-
    velopment (OERD) 8
Ontario Building Code 148

Ontario Facilities Energy Con-
    sumption Directive 162
Ontario Power Authority (OPA)
    133
Ontario Regulation (O. Reg.)
    397/11 172
Ontario's IESO 249
open collaborative model 262

P
Pan-Canadian Framework on
    Clean Growth and Climate
    Change 11, 129, 222
Paris Agreement on Climate
    Change (COP21) 66
peak demand 116, 134, 136, 143,
    150, 156, 187
PEI Energy Corporation 179
PEI Energy Strategy 2017 180
PEI Office of Energy Efficiency
    177
PEI Provincial Energy Strategy,
    2016/17 178
PNS energy savings 113, 117
policy recommendations 239
POWER.HOUSE program 142
Power Smart Solar Energy Pro-
    gram 61
process evaluation 242
Process Integration Incentive Pro-
    gram 20
program incentives 120
Property Assessed Clean Energy
    (PACE) 120
province's Climate Action Charter
    41
PST exemption 217
public purpose energy service
    companies (PPESCOs) 234

Q

Quebec Energy Strategy 2006-2015: Using Energy to Build the Quebec of Tomorrow 188

Quebec Energy Utility Investments 205

Quebec GHG Emission Reduction Targets 199

Qulliq Energy Corporation (QEC) 127

R

R-2000 homes 211

R-2000 label 18

R-2000 program 255

R-2000 Standard program 4

Régie de l'énergie du Québec 192

regional planning 146

regulations 15

Residential Energy Efficiency Program (REEP) 81

Residential No-Charge Energy Savings Program 26

retrofit 72

Retrofit program 136, 154

retrofits 44, 45, 165, 267, 271

RETScreen® 12, 248

Expert 268

S

Saskatchewan Conservation House 211

Saskatoon Light and Power (SL&P) 213

SaskEnergy 213

SaskPower 213, 237

shareholder incentive 171

SMART Green 165

social benchmarking 142

SOE Training and Support incentives 155

Solar Power Demonstration Site 214

Specified Gas Emitter Regulation (SGER) 29

Steering Committee on Energy Efficiency (SCEE) 8

Strategic Evaluation Division 19

strategy plan 125

Summerside Electric 179

supply side 54

Sustainable Building Manitoba 56

T

takeCHARGE 78

partnership 80

Temperance Street Passive House 213

Territorial Power Subsidy Policy 101

Territorial Power Subsidy Program 97

The Atmospheric Fund (TAF) 163, 236

Toronto Green Standard 272

Toronto's 2050 climate target 272

Total Resource Cost (TRC) 38

test 158

TowerWise 271

program 236

Transition Énergétique Québec 191

Transport Canada 9

U

UARB 117

Utilities Commission Act 42

Utility and Review Board 260
Utility Educational Programs 88
utility on-bill financing 89

V
Vancouver Building Bylaw 45
Vancouver Charter and Commu-
    nity Charter 44
Vancouver Declaration on Clean
    Growth and Climate Change
    11
VerEco Demonstration Home 212

W
Western Climate Initiative (WCI)
    carbon market 203

Western Development Museum
    (WDM) 214
White Paper, New Brunswick En-
    ergy Policy 2000-2010 65
wholesale market 133, 145

Y
Yukon Government Energy Solu-
    tions Centre 223
Yukon Government Energy Strat-
    egy 222
Yukon Housing 222
Yukon Utilities Board 223